# THE REVIVAL OF POLITICAL IMAGINATION

# THE REVIVAL OF POLITICAL IMAGINATION

## UTOPIAS AS METHODOLOGY

*Edited by Teppo Eskelinen*

BLOOMSBURY ACADEMIC
LONDON • NEW YORK • OXFORD • NEW DELHI • SYDNEY

BLOOMSBURY ACADEMIC
Bloomsbury Publishing Plc
50 Bedford Square, London, WC1B 3DP, UK
1385 Broadway, New York, NY 10018, USA
29 Earlsfort Terrace, Dublin 2, Ireland

BLOOMSBURY, BLOOMSBURY ACADEMIC and the Diana logo
are trademarks of Bloomsbury Publishing Plc

First published in Great Britain 2020 by ZED BOOKS
This paperback edition published by Bloomsbury Academic 2022

Cover design: Burgess & Beech

A catalogue record for this book is available from the British Library.

A catalog record for this book is available from the Library of Congress.

ISBN: HB: 978-1-7869-9959-7
PB: 978-0-7556-4994-5
ePDF: 978-1-7869-9958-0
eBook: 978-1-7869-9956-6
mobi: 978-1-7869-9957-3

Typeset by Swales & Willis Ltd.

To find out more about our authors and books visit
www.bloomsbury.com and sign up for our newsletters.

# CONTENTS

# NOTES ON CONTRIBUTORS

**Inkeri Aula** (Ph.L) is a cultural researcher living in a utopian community in the Finnish archipelago. She currently works as a lecturer in cultural anthropology in the University of Eastern Finland UEF and in the ERC-funded transdisciplinary research project SENSOTRA. Her research interests include Afro-Brazilian culture, transnational capoeira, environmental relationships, sensory mediations and transgenerational ethnography.

**Teppo Eskelinen** (Dr.Soc.Sc) is Senior Lecturer in Development Studies at the University of Jyväskylä, Finland. A philosopher by education, Eskelinen has also worked as a lecturer in social and public policy. His research interests include political economy and global justice broadly understood. He has also done research on radical democracy, classics of heterodox economics and economic alternatives.

**Jarno Hietalahti** (Ph.D.) is a postdoctoral researcher at the University of Jyväskylä, Finland. His main field of interest is the philosophy of humor and humanistic thinking, especially Erich Fromm's radical humanism. Hietalahti has published over 20 peer-reviewed and non-peer-reviewed scientific articles on humor. He is also the author of the first Finnish non-fiction book (peer-reviewed) on the philosophy of humor and laughter, *Huumorin ja naurun filosofia* (2018).

**Maria Laakso** (Ph.D.) is a postdoctoral researcher in literary studies at Tampere University, Finland. She specializes in narrative theory, literary narratology and societal themes in contemporary literature. Her current research interests include dystopian and apocalyptic narratives, and imagining/narrating the future in contemporary literature and culture.

**Keijo Lakkala** (M.Soc.Sci) is a doctoral student at the University of Jyväskylä, Finland. He is interested in radical traditions within political philosophy and especially how these traditions relate to the concept of utopia.

**Aleksi Lohtaja** (M.Soc.Sci) is a doctoral student in political science at the University of Jyväskylä, Finland. His Ph.D. research focuses on the intersections of architectural utopias, cultural politics and political philosophy.

**Olli-Pekka Moisio** (Dr.Soc.Sci) is Senior Lecturer of Philosophy in the Department of Social Sciences and Philosophy, University of Jyväskylä, Finland.

**Miikka Pyykkönen** (Dr.Soc.Sci) is Professor of Cultural Policy at the University of Jyväskylä, Finland and Docent of Sociology at the University of Helsinki. His current research interests include cultural policy, creative economy and entrepreneurship, civil society and government, and histories of the governance of ethnic minorities. He is currently translating Professor Erik Olin Wright's book *Envisioning Real Utopias* into Finnish with the research team on utopian thinking.

**Matti Rautiainen** (Ph.D.) is Senior Lecturer in Pedagogy and Social Sciences in the Department of Teacher Education at the University of Jyväskylä, Finland. His research interests are pedagogy of history, education for democracy and teacher education.

# ACKNOWLEDGMENTS

Most of the work on this book was enabled by a generous grant from the Kone Foundation, for the project 'Utopia as Method: Articulations of Alternative Realities'. In addition, Maria Laakso's article was completed in the context of another project funded by the Kone Foundation, 'Darkening Visions: Dystopian Fiction in Contemporary Finnish Literature', and Inkeri Aula's article is an outcome of the projects 'Transforming the Future of Brazil: Ritual and Indigenous Agencies' (University of Helsinki research funds) and SENSOTRA (ERC advanced grant #694893).

Hanna Kuusela and Mikko Jakonen were initially part of the writing group. While both chose to withdraw from the process, their input to the collective thinking behind the book is appreciated. Thank you also Marcus Petz for the English-language revision. Finally, we would also like to extend a general appreciation to everyone who commented on the articles and interacted with us during the work.

PART I

**DIAGNOSIS**

# 1 | INTRODUCTION: UTOPIAS AND THE REVIVAL OF IMAGINATION

*Teppo Eskelinen, Keijo Lakkala and Maria Laakso*

Societal utopias have so frequently been pronounced to be dead, that declaring their demise has become something of a cliché. The general mentality of the 1990s led to the majority of laymen and intellectuals alike to assume this assertion as common sense. According to a widely assumed idea, utopias became obsolete as humanity ascended from the era of totalitarian ideologies to the era of liberal capitalist democracies. In this new order characterized by liberty, no human being is forced to adapt to any grand utopian vision imposed by others. Utopias have a long history, for sure – from antiquity on, with the term coined by Thomas More in 1516. But in our era, society would be different.

In this context, it is difficult to avoid mentioning Francis Fukuyama. In 1989, the then deputy director of the US State Department's policy planning staff published an article which achieved symbolic status: 'The End of History?' Based on Alexandre Kojève's reading of Hegel's philosophy of history and inspired by the collapse of the Soviet bloc, the article proclaimed that history in the sense of fundamental contradictions was over (Fukuyama 1989, 8).

Yet the novelty of Fukuyama's ideas should not be overstated. The exhaustion of utopian energies and political ideas was noted already three decades earlier by Daniel M. Bell (1960). In political theory texts published in the 1980s, the anti-utopian 'postmodern' sentiment of late stage capitalist societies was quite visible. Jürgen Habermas in 'The New Obscurity' (1986) argued, that the modern time consciousness based on the ideas of progress and revolution had become narrower, along with the horizon of the future (Habermas 1986, 2). Jean-François Lyotard analyzed postmodernism as the intellectual condition in which metanarratives of modernism, including utopianism, have been left behind (Lyotard 1984). Two years before Fukuyama's article, Krishan Kumar already asked: 'Can there be anything more commonplace than the pronouncement that, in the twentieth century, utopia is dead – and beyond any hope of resurrection?' (Kumar 1987, 30).

Fukuyama's position might be simplifying to the point of ridiculousness, but it signals a sign of the times. This is the societal condition in which it is almost impossible to think of alternatives to liberal democratic capitalism. To abridge Slavoj Žižek (2009, 53):

> Though it is easy to make fun of Fukuyama's notion of the End of History, the majority today is Fukuyamaist. Liberal-democratic capitalism is accepted as the finally found formula of the best possible society; all one can do is to render it more just, tolerant and so on.

A pressing question given this condition is, then, what does the belief in this 'finally found formula' do to key political skills, such as reflection, public criticism and political imagination? Can politics retain a sense of meaning if we believe that it has, more or less, arrived at its destination? Any plausible answer will be hardly short of terrifying. An immediate question then arises: Can the skill to imagine other kinds of societies be saved; and if it can, how?

## The Specificity of Absolutist Utopias

It is natural to begin the rehabilitation of utopian thought by analyzing the intellectual mechanisms feeding into its demise. Indeed, the withering away of utopias is not only a common mentality, but the subject (and partially also an outcome) of thorough theorizing. There is no shortage of sophisticated sociology warning us about the dangers of utopian thought. Most importantly, the popular anti-utopian sentiment discussed above is based on the interpretation of utopias as necessarily *absolutist utopias*. This interpretation is a very particular one, and hardly exhaustive. It approaches utopias as static models, the implementation of which can only take place in the manner of imposing a blueprint upon society. On this basis, it is easy to interpret utopias as signifying nothing but totalitarianism, as opposed to liberty.

Absolutist interpretations of utopias can take a variety of forms. Aspects of absolutist utopias emphasized by different authors are moral monism, holistic methodology and utopias as closed systems.

The critique of utopian thought as moral monism, as opposed to plurality and versatility, is particularly associated with Isaiah Berlin (Berlin 1997a,b). In Berlin's (1997a, 5) words, utopias assume an objective and coherent, and unavoidably dogmatic, system of 'moral truths'. In such a system, for every genuine moral question there can be only one correct answer, there is a reliable method for finding the correct moral answers, and all correct answers to moral questions are compatible with each other (Berlin 1997a, 5). Yet a perfect whole, 'the ultimate solution' (Berlin 1997a, 11) to moral questions, or a 'perfect social harmony' (Berlin 1997b, 191), are conceptually incoherent ideas. A choice between different values is always necessary. 'We are doomed to choose, and every choice may entail an irreparable loss' (Berlin 1997a, 11). No society can realize all values coherently. This, according to Berlin, renders utopias impossible.

The interpretation of utopias as based on a holistic methodology leading to totalitarianism derives chiefly from the work of Karl Popper. According to Popper, the utopian desire for impossible perfections and this striving for perfection will inevitably cause violence and repression. For Popper, utopianism is a view according to which 'rational political action must be based upon a more or less clear and detailed description or blueprint of our ideal state, and also upon a plan or blueprint of the historical path that leads towards this goal' (Popper 1963, 358). The concept of utopia is thereby associated with social blueprints and further with totalitarianism (e.g. Schapiro 1972, 85; Popper 1963, 357–360). The rational organization of the ideal society that Popper calls *utopian engineering* is inevitably in the hands of few and therefore inclined to violence and totalitarian control (Popper 1963).

A third interpretation in absolutist fashion is to see utopias as closed and static systems. Ralf Dahrendorf (1958, 116) argued, that one structural characteristic of utopias is their uniformity, based on a universal consensus on values and institutional arrangements, and the absence of disagreement and conflict.

Utopias are perfect – be it perfectly agreeable or perfectly disagreeable – and consequently there is nothing to quarrel about. Strikes and revolutions are as conspicuously absent from utopian societies as are parliaments in which organized groups advance their conflicting claims for power. (Dahrendorf 1958, 116)

Utopias might have 'a nebulous past' (Dahrendorf 1958, 116), but they do not have a future. Utopias 'are suddenly there, and there to stay, suspended in mid-time or, rather, somewhere beyond the ordinary notions of time' (Dahrendorf 1958, 116).

Many other theorists have landed on interpretations similar to those of these canonical authors. John Gray in *Black Mass: Apocalyptic Religion and the Death of Utopia* (2007, 2) claims, that the whole of Western history has been terrorized by utopian projects: 'entire societies have been destroyed and the world changed forever'. J. L. Talmon (1952, 252) defines utopia as the 'complete harmony of interests, sustained without any resort to force, although brought about by force'. Hans Achterhuis argues, that as utopias are seen as perfections, utopia sees itself as legitimizing all violence that could potentially be needed in this realization (Achterhuis 2002, 160–161). All in all, the intellectual discourse on utopian thought has been dominated by thinkers who 'describe utopianism as a one-way ticket to totalitarianism' (Oudenamspen 2016, 43).

This strong association of utopias with totalitarianism can be understood as an intellectual reflection of the traumas of the twentieth century; 'mankind's darkest hour so far'. The rise of totalitarian regimes, two World Wars, atomic bombings, the Holocaust and the Cold War have left the world in a state of 'cosmic pessimism', to paraphrase Krishan Kumar (1987, 380). When imagining alternative societies, quite a few people have in mind something along the lines of George Orwell's *Nineteen Eighty-Four* (1949, 390): 'If you want a picture of the future, imagine a boot stamping on a human face — forever'.

Yet this fear has already become clearly overstated. The accounts of utopias discussed above have, for a good reason, been called 'dystopic liberalism' (Thaler 2018), or works by 'liberalists of fear' (Shklar 1989), in reference to their systematic preoccupation (or should we say obsession) with political evil. Saving humanity from totalitarianism has become an obstacle to political progress and imagination, a justification of whatever is wrong in the current society as a lesser evil by definition. Dystopic liberalism allows little role for political progress, limiting it to 'piecemeal social engineering'.

Indeed the absolutist approach to utopias, while widely assumed in popular discourse, is far from being the whole truth about utopias. The absolutist position can be contrasted with the 'relationalist' interpretation. Relationalism sees utopias as first and foremost criticisms and counter-images of the present. They do not exist to be attained as such, informing a holistic description of how society should be constructed, but to show that alternatives to the present exist. Like Terry Eagleton (2009, 33) phrases, the alternative worlds in utopian fiction are devices for 'embarrassing the world we actually have'. In fact, it is very much possible to think of utopia as an epistemological rather than ontological category. Utopian texts can be understood as heuristic tools for social imagination rather than 'architectural' blueprints for an ideal society. We are interested here exactly in this 'critical function' of utopias, or the role of utopias in criticizing and relativizing the present by showing a radical alternative to it.

The concept of 'utopia', as such, does not carry any absolutist connotations. As a combination of Greek *topos* (for place), and (depending on the interpretation) *ou* for general negative or *eu* for good (or ideal and prosperous) (see Manuel and Manuel 1979, 1), '(e)utopia' carries the double meaning of no-place/good-place. 'A good place that does not exist' can well be a dream or a criticism, rather than a blueprint.

Because the present is never static, utopian counter-images assume different interpretations at every point in time. The relationship between reality and dynamic utopias therefore constantly changes, rather than reality just approaching a given fixed utopia. Utopias force the readers to reflect on their own *topos*. In its most general interpretation, the concept of utopia refers to a place that is more desirable than the one we currently inhabit (Suvin 1997, 126–128). For the relationalist, the idea that utopias are about perfection is a crude misunderstanding (see for example Abensour 2008; Claeys 2017).

With the alarm about utopias as totalitarian, we risk losing utopias as tools for reflection and imagination. The loss of this critical function with the demise of utopias is an alarming symptom of our time. What is at stake is not only critical reflection on society, but the very human capability to conceive of a different world. Part of the modern degradation of utopianism has indeed been 'an incremental impoverishment of what might be called Western imagination' (Jacoby 2005, 5). As totalitarian ideas collapsed under the victory of liberal capitalism, we were made to believe that not only communism, but by and large the quest for a better society was over. The association of utopias with totalitarianism then enforces the idea, that there are no existing or even conceivable alternatives to global capitalism (Levitas and Sargisson 2003, 15). As noted above in the quote by Žižek, liberal capitalism is taken as a given, and the hope invested in politics is based on an attempt to qualitatively improve this given system. It can be made 'more just or more tolerant', but no systemic change is possible. This condition is hardly assisted by the current ecological state of emergency, which might make mere survival seem like a hopeful scenario, and a qualitatively better society beyond the horizon of possibility.

The point of saving political imagination by fostering utopias also has implications on the relation between utopias and their ostensible opposites. Namely, the opposing counterparts of utopias are typically seen to be dystopias. The word 'dystopia' (a neologism from Greek *dis topos* – bad place) was used for the first time by John Stuart Mill in 1868 in a speech criticizing government's policy on Irish property (Milner 2009, 827), and today it is widely used for describing the idea of utopia gone astray. But while literary dystopias clearly describe unpleasant realities, instead of being the opposite of utopias, they can well turn out to serve the same function in terms of critical reflection.

Dystopian and utopian imaginaries alike then function as criticisms of the existing society. Both can be interpreted as *social dreaming* (Sargent 1994), the key point of which is to support the development of imagination through fiction (see Laakso in this volume). Utopias can be dreams of distant places or dreams of the future, but they are essentially social dreams all the same. Similarly, dystopias show alternative worlds for reflecting upon the existing one. Therefore, the opposite of utopias and dystopias alike is rather the inability to dream.

### Towards a Methodological Understanding of Utopianism

The quest for the revival of utopian imagination and uncovering the functions of utopias in totality requires going back to their roots. The discourse and critical use of utopianism is, in any case, a long tradition in Western political and philosophical thought. Utopianism as social dreaming has traditionally had three important manifestations (Sargent 2010, 5; 1994, 2–3): literary utopias (including non-literary narrative and/or fictional utopias like drama), utopian practice (for example experiments in communal living), and social theory. Utopias then vary significantly in form depending on whether we are talking about utopia as a literary genre, a set of counter-practices, or as a specific form of political thought.

As we approach utopias as a method for fostering political imagination and critical thinking, all these manifestations are present. Further, utopias need to be understood in a strongly relational sense. Even if a utopian society is described in the form of a closed system, it can be approached as thematically open, since it always relates itself to both the society in which it was written and the society in which it is read. Indeed utopia philosophically means more a method for thinking societal alternatives than it means ontological space.

Further, even when taking distance from utopia (as a noun) it is necessary to focus on the utopian (as an adjective).[1] Utopianism (or utopian mentality) should be understood as a general orientation or a quality, as opposed to comprehensive imaginary reconstitutions of society (see Levitas 2013). Anything that expresses an orientation or a desire towards a qualitatively better mode of being, can be perceived as utopian. In other words, 'utopia' refers to 'a non-existent society described in considerable detail and normally located in time and space', while 'utopianism' refers to 'social dreaming' (Sargent 1994, 3–7), as noted above. In the quest for reviving political imagination, the most relevant objects of study will rather relate to utopianism, as we seek utopianism in a large number of existing spaces and practices. The main task is then not the reconstruction of any imaginary society, but locating utopian hope in the world around us.

Human beings constantly orient themselves towards a better state of being. For Karl Mannheim, utopian mentality is something that orients towards a new *topos* from the present, transcends reality and 'breaks the bonds of the existing order' (Mannheim 1979, 173). Leszek Kolakowski defined utopia as 'a state of social consciousness, a mental counterpart to the social movement for radical change in the world' (Kolakowski 1968, 69). In Ruth Levitas' (2010) words, utopias express the 'desire for a better being'. Frank E. Manuel and Fritzie P. Manuel (1979, 5) called this desire the 'utopian propensity' which has manifested itself in diverse forms of human experience throughout the history of mankind. This desire, propensity, or in Ernst Bloch's terms 'principle of Hope' (Bloch 1986) can be seen as a core element of any utopia. According to Vincent Geoghegan (2008, 17),

> we can speak of a utopian disposition, a utopian impulse or mentality, of which the classic utopia is but one manifestation. This impulse is grounded in the human capacity, and need, for fantasy; the perpetual conscious and unconscious rearranging of reality and one's place in it. It is the attempt to create an environment in which one is truly at ease.

However, there is a given difference between Bloch's 'principle of Hope' and Levitas' 'desire'. According to Levitas, the concept of hope does not include all forms of utopia. Hope would be in vain if there would not be real prospects of the realization of its object. The concept of desire, however, is less strongly connected to explicit utopias. Human beings can desire even something completely impossible or have vaguely directed desires. One can desire a better being even though there is no hope of this desire being fulfilled. In some cases, there is of course, hope for the fulfilment of desire, but desire itself is not the same as what is hoped for. The extension of the concept of hope is then narrower than that of desire; yet both are motivating factors for the search for a better society that need to be taken into account.

Another useful distinction on the different ways to understand utopias as method, can be found in Levitas' *Utopia as Method: The Imaginary Reconstitution of Society* (2013). Levitas distinguishes analytically three methodological (often intertwined) modes that utopia can take: the architectural, the archeological and the ontological. Of these three, the *architectural* mode refers to depictions of a better world, designs of a better society and delineation of the good society, a more or less detailed picture of a desired world. This 'imagining a reconstructed world and describing its social institutions' (Levitas 2013, 197) is the understanding of utopias on which most critics of utopianism base their skepticism.

The approaches closer to our starting point are the archeological and ontological approaches. The *archeological* mode of utopian method involves identifying utopian elements in what is typically seen as pragmatic or non-utopian. It pieces together utopian elements embedded for example in political programs and social and economic policies (Levitas 2013, 153). This mode comes close to the so-called 'utopian hermeneutics'. This interpretative research orientation aims to find utopian elements in all areas of human culture. This kind of orientation can be found for example in the works of Ernst Bloch (1986), Douglas Kellner (1997) and Fredric Jameson (1979). Especially in Bloch's *The Principle of Hope* (1986), utopian archeology (or utopian hermeneutics) finds utopian, premonitory and prefigurative images of the future from the works of the past and catalogues this utopian surplus from the early Greek philosophers to the present day. Further, the *ontological* mode of utopian methods means imagining ourselves otherwise. It also entails a judgment about what constitutes human flourishing. According to Levitas (2013, 196), the central point of the ontological mode of utopian method 'is that the utopian method necessarily involves claims about who we are and who we might and should be'.

Here, we want to depart from the architectural mode and focus on the latter two. The task is to understand how utopias can be used, how they affect human beings, and most importantly, to find utopian sentiments in various kinds of spaces and places. Utopian elements can be included in political initiatives which are very much part of the everyday political discourse, such as Universal Basic Income (UBI), for instance (e.g. Bregman 2016). Or they can be found in experimental politics, contentious spaces or even ostensibly self-evident political concepts. The methodological challenge is to recognize utopian propensities and their relevance when observing them.

## Utopianism Facilitating Transformation

Yet the point of utopianism is not only to foster imagination. Utopias exist also to facilitate social transformation, even if not in the absolutist sense. Utopia is then not only a method of reflection, but also a method of changing the world. This approach comes close to what Suman Gupta (2001) has called 'rational utopian thinking', being essentially about revitalizing 'an effective political will'. Indeed, political imagination cannot be detached from movements for changing the world.

Another way of interpreting utopia as a method for social transformation is to define utopia as a counter-image or 'a compass'.

As a counter-image, utopia reflects social problems of its time and generally relativizes the present. The function of utopias as counter-images emphasizes their dynamic and relational nature. In the sense of 'a compass', utopias can also be seen to provide a direction (and a purpose) for societal change. Erik Olin Wright's (2010) metaphor of the 'socialist compass' highlights the idea that the function of utopias is to provide an idea of the direction into which society should evolve. The metaphor further implies that on the way to a better society, various kinds of detours will be experienced, yet awareness of the general direction remains.

Utopias as counter-images can also be developed into counter-practices that concretely enable doing otherwise here and now (on the connection between counter-image and counter-practice, see Lakkala in this volume). In the age of 'capitalist realism' (Fisher 2009), the main function of utopia is not to articulate goals for political programs or any detailed visions, but to open the present for the possibility of a better being, so that systemic alternatives become visible again.

This goal can be expressed with the concept of the facilitating function of utopias, as opposed to their critical function. The function of facilitating social change takes place by creating hope and presenting goals for transformative social action. The critical function, on the other hand, enables questioning and criticizing the present society (Levitas 2010), but is not directly relevant for transformative movements. Utopias are always critical, yet not necessarily facilitating.

Even within relatively coherent theoretical approaches utopias have been interpreted both as facilitating social change and as functioning as an obstacle to it. For instance within Marxism, Marx and Engels (2004 [1848]) fiercely criticized utopian socialism for preventing social change (Levitas 2010, 6).[2] According to them, utopian socialism 'formed mere revolutionary sects' and built 'castles in the air'. According to their point of view, utopian socialists, while progressive in their own time, became unable to facilitate social movements, as the dynamism of struggle had changed with the passing of time. In a more analytical fashion, Ernst Bloch (1986, 205) made a distinction between the 'cold and warm streams' of Marxism. The cold stream of Marxism, as its critical-scientific manifestation, refers to the conditional analysis of the whole historical situation, unmasking of ideologies and disenchantment of meta-physical illusions. The warm stream of Marxism, on the other hand, is the facilitating and mobilizing function, containing the revolutionary enthusiastic emancipatory intention and the societal (socialist) vision.

In his 1967 lecture delivered at the Free University of West Berlin, Marxist philosopher Herbert Marcuse described this function of utopias as follows:

> Utopia is a historical concept. It refers to projects for social change that are considered impossible. Impossible for what reasons? In the usual discussion of utopia the impossibility of realizing the project of a new society exists when the subjective and objective factors of a given social situation stand in the way of the transformation – the so-called immaturity of the social situation. Communistic projects during the French Revolution and, perhaps, socialism in the most highly developed capitalist countries are both examples of a real or alleged absence of the subjective and objective factors that seem to make realization impossible. (Marcuse 2014, 250)

The function of preventing societal change can be found mainly in the abstract and static forms of utopian thought. Dreaming of abstract utopian worlds can sometimes become an end in itself, negating concrete action. Lewis Mumford (1922, 20) called such abstract utopias 'utopias of escape'. According to Mumford (1922, 20), a utopia of escape is 'an enchanted island' where one loses the 'capacity for dealing with things as they are'. In contrast, Mumford introduced the notion of the 'utopia of reconstruction'. A utopia of escape leads back to the ego of the utopian thinker, while the utopia of reconstruction orientates towards the outside world and aims to change it. In this book, we strongly emphasize an approach based on the notion of utopia of reconstruction: Utopias exist for social transformation to be possible.

## The Purpose of This Book and the Content of the Chapters

To be sure, this is not the first contemporary book written on uto-pias, not even on their methodological aspects (e.g. Levitas 2013, 2010; Wright 2010; Chrostowska and Ingram 2017; Sargisson 2012;

Žižek and Thompson 2013). Several authors have noted the need to recognize the open, dynamic and reflexive nature of utopias, and generally the need for utopian thought and horizons beyond the existing (liberal capitalist) social order. Our emphasis is on hope and imagination, and their value for transformative purposes. Utopianism both requires and fosters an imaginative skill. This skill needs to be practiced if it is to thrive, be challenged in transformative projects and movements, and most importantly be practiced collectively. While imagination might sound like private and directionless daydreaming, transformative imagination is a collective and reflexive skill. Therefore, we want to depart from philosophical accounts of utopias focusing on conceptual distinctions, and focus on this utopian skill.

This analysis requires using a number of highly diverse examples. Perhaps utopianism is best fostered in counter-practices, as they properly relate to the critical function of utopias. Maybe we need to look at existing expressions of private hope and ask, how could it be collectivized? Perhaps fostering utopian skills is brought about by reflection through dystopian literature, seeing the utopian element in education, or experimental humor. Or maybe we need to study repressed but (proudly) organized social spaces, architecture and the city as manifestations of utopianism, or ostensibly commonplace political concepts. The numerous case studies in the book develop these different approaches and topics.

In this spirit, we want also to emphasize a departure from the tradition of 'utopian studies', in the sense of locating literary or real utopias and then analyzing them. Our concrete aim is to produce social science for social transformation, and utopias function as vital inspiring and reflexive tools for this purpose. In the spirit of the notion of 'utopia as method', we do not wish to study utopias, but to use them. Utopias are thereby seen primarily as functional, rather than as literary products.

The book continues with a theoretical input, analyzing the relation between utopias as counter-images and utopias as counter-practices, to reflect on the question of the facilitating potential of utopianism. Written by Keijo Lakkala, Chapter 2 develops further the notion of utopias as critical counter-images and attempts to restate the essence of utopias as critical and facilitating. The chapter shows that utopias can still have dynamic qualities, and be essentially open to the future. The critical function of utopias is further developed in the context of critical counter-practice, in which 'cracks' are actively sought in the existing social order.

The third chapter continues by analyzing the state of utopias today as privatized hope. As noted above, a given desire for a better mode of being is ubiquitous in humanity, yet it can appear in a variety of different forms. Teppo Eskelinen, Keijo Lakkala and Miikka Pyykkönen ask, how this hope has become individualized and backward-looking. This is assisted by an analysis of how hope is invested in self-control fantasies of entrepreneurship and individual escapes and on how the collective imagination turns into a nostalgic mood. The chapter calls for a recollectivization of hope, and gives some ideas for tools for this recollectivization.

These more general and diagnostic chapters are followed by case studies. First, based on the development of the concept of the 'anthropology of hope', Inkeri Aula shows how apparently destitute places and spaces can have a utopian orientation – indeed part of the methodological approach of the anthropologist should be to discern and understand such hope, and further to assist in developing utopian aspirations grounded in this hope. Through a case study of *quilombo* communities in Brazil with rich observation data, Aula shows how transformative inspiration can be maintained and built.

In the subsequent chapter, Maria Laakso moves to a case of a very different kind, the dystopian novel. Using Margaret Atwood's *The Handmaid's Tale* as reference point, Laakso shows the relevance of dystopias for social dreaming and thereby political imagination. Further, the article discusses the function of narratives: The narrative nature of the literary novel allows the reader to assume the perspective of a resident of an imaginary social reality. The chapter shows that the assets of dystopian fiction to imagination and critical reflection do not depend even on the time of the writing of the dystopia.

The chapter by Olli-Pekka Moisio and Matti Rautiainen takes up yet again a quite different kind of case: education. Education materializes kinds of utopian hopes, articulated in curricula, and is generally an attempt to prepare people for a world that does not yet exist. On this basis, Moisio and Rautiainen discuss education as enabling human realization. Reflecting on a case study of a Finnish rural school, they show how a school can be informed by hope and a transformative spirit, if tools exist for collective reflection. The ambitions in education can be high, but are rarely achievable by the lone educator.

After these case studies, the book turns again towards a more theoretical mode. Jarno Hietalahti's chapter discusses a necessary yet largely overlooked aspect of utopian thought, which is humor. The importance of the topic derives not only from the misunderstanding of utopias as necessarily perfect and thereby dull, but also from the significance of experimentation. Looking first at the functioning of conservative laughter *at utopia* and the radical laughter *with utopia*, Hietalahti proceeds to seek to transcend this distinction with the notion of laughter *in utopia*, developing an idea of humor functioning on language in which the syntax does not yet make sense to us, being scary and disturbing and thereby opening new social horizons.

Aleksi Lohtaja's chapter continues by discussing architectural utopias and more generally the question of the possibility of utopia in architecture. Seen typically as non-utopian and merely material, architecture provides experimental spaces which can also manifest ideas related to transformative aspirations. Comparing accounts on the relationship between architecture, ideology and utopias, Lohtaja shows how utopian ideas might exist in city-planning projects, where one does not necessarily look for them.

In the final chapter of the book, Teppo Eskelinen discusses democracy and utopias. The point of the piece is to contribute to the discussion on utopianism by showing how an ostensibly commonplace concept such as democracy can entail utopian elements. This requires an understanding on how radical concepts are diluted and stripped of utopian quality within the existing hegemonic framework of political thought. Democracy remains to be transformative, even revolutionary, if understood consistently and without the restrictions currently imposed on it. The author then sketches the starting points of utopian democracy, discusses its current limitations, and suggests ways forward.

Through these numerous case studies and theoretical inputs, the book makes open and tacit contributions towards its purpose: the revival of political imagination through utopianism. Seeing horizons of collective change is indeed a skill, and fostering this skill depends on whether one is tuned to see glimpses of political hope, whether people can come to together to conceive of change and experiment with it, and whether we are critically aware of the background of our current social predicament. Varied though the perspectives within this book are, this heterogeneity is part of the experimental approach needed on the way to the revival of the utopian spirit.

## Notes

1 Another way to illuminate this distinction is to contrast utopian programs (realization of a new totality) and the utopian impulse found in political theories, philosophy and cultural products (see Jameson 2005, 6).

2 In a stricter interpretation, Lenin (2004) saw all utopias as harmful because of their necessary detachment from the political class forces: 'In politics utopia is a wish that can never come true – neither now nor afterwards, a wish that is not based on social forces and is not supported by the growth and development of political class forces'.

## References

Abensour, M. (2008) 'Persistent Utopia'. *Constellations* 15 (3), 406–421.

Achterhuis, H. (2002) 'Violent Utopias'. *Peace Review* 14 (2), 157–164.

Bell, D. (1960) *The End of Ideology: On the Exhaustion of Political Ideas in the Fifties.* Cambridge, MA and London: Harvard University Press.

Berlin, I. (1997a) 'The Pursuit of the Ideal'. In H. Hardy and R. Hausheer (eds.) *The Proper Study of Mankind: An Anthology of Essays.* London: Chatto & Windus, 1–16.

Berlin, I. (1997b) 'Two Concepts of Liberty'. In H. Hardy and R. Hausheer (eds.) *The Proper Study of Mankind: An Anthology of Essays.* London: Chatto & Windus, 191–242.

Bloch, E. (1986) *The Principle of Hope*, Volume 1. London: Blackwell.

Bregman, R. (2016) *Utopia for Realists: The Case for a Universal Basic Income, Open Borders and a 15-Hour Workweek.* Amsterdam: The Correspondent.

Chrostowska, S. D. and Ingram, J. D. (eds.) (2017) *Political Uses of Utopia: New Marxist, Anarchist and Radical Democratic Perspectives.* New York: Columbia University Press.

Claeys, G. (2017) 'When Does Utopianism Produce Dystopia?' In Z. Cziganyik (ed.) *Utopian Horizons: Ideology, Politics, Literature.* Budapest and New York: Central European University Press, 41–62.

Dahrendorf, R. (1958) 'Out of Utopia: Toward a Reorientation of Sociological Analysis'. *The American Journal of Sociology* 64 (2), 115–127.

Eagleton, T. (2009) 'Utopia and Its Opposites'. In L. Panitchs and C. Leys (eds.) *Socialist Register 2000: Necessary and Unnecessary Utopias. Social Register,* Volume 36, 31–40.

Fisher, M. (2009) *Capitalist Realism: Is There No Alternative?* Winchester and Washington, DC: O Books.

Fukuyama, F. (1989) 'The End of History?' *The National Interest* 16, 3–18.

Geoghegan, V. (2008) *Utopianism and Marxism.* Bern: Peter Lang.

Gray, J. (2007) *Black Mass: Apocalyptic Religion and the Death of Utopia.* London: Allen Lane.

Gupta, S. (2001) *Corporate Capitalism and Political Philosophy*. London: Pluto Press.

Habermas, J. (1986) 'The New Obscurity: The Crisis of the Welfare State and the Exhaustion of Utopian Energies'. *Philosophy and Social Criticism* 11 (2), 1–18.

Jacoby, R. (2005) *Picture Imperfect: Utopian Thought for an Anti-Utopian Age*. New York: Columbia University Press.

Jameson, F. (1979) 'Reification and Utopia in Mass Culture'. *Social Text* 1, 130–148.

Jameson, F. (2005) *Archeologies of the Future: The Desire Called Utopia and Other Science Fictions*. London and New York: Verso.

Kellner, D. (1997) 'Ernst Bloch, Utopia, and Ideology Critique'. In J. O. Daniel and T. Moylan (eds.) *Not Yet: Reconsidering Ernst Bloch*. London and New York: Verso, 80–95.

Kolakowski, L. (1968) 'The Concept of the Left'. In L. Kolakowski: *Towards a Marxist Humanism: Essays on the Left Today*. New York: Grove Press.

Kumar, K. (1987) *Utopia and Anti-Utopia in Modern Times*. Oxford: Basil Blackwell.

Lenin, V. I. (2004) 'Two Utopias'. In *Lenin Collected Works*, Volume 18. Marxists Internet Archive. Retrieved from www.marxists.org/archive/lenin/works/1912/oct/oo.htm.

Levitas, R. (2010) *The Concept of Utopia*. Bern: Peter Lang.

Levitas, R. (2013) *Utopia as Method: The Imaginary Reconstitution of Society*. New York: Palgrave Macmillan.

Levitas, R. and Sargisson, L. (2003) 'Utopia in Dark Times: Optimism/Pessimism and Utopia/Dystopia'. In T. Moylan and R. Baccolini (eds.) *Dark Horizons: Science Fiction and the Dystopian Imagination*. New York and London: Routledge, 13–28.

Lyotard, J.-F. (1984) *The Postmodern Condition: A Report on Knowledge*. Minneapolis, MN: University of Minnesota Press.

Mannheim, K. (1979) *Ideology and Utopia: An Introduction to the Sociology of Knowledge*. London and Henley: Routledge & Kegan Paul.

Manuel, F. E. and Manuel F. P. (1979) *Utopian Thought in the Western World*. Oxford: Basil Blackwell.

Marcuse, H. (2014) 'The End of Utopia'. In D. Kellner and C. Pierce (eds.) *Marxism, Revolution and Utopia: Collected Papers of Herbert Marcuse*. London and New York: Routledge, 249–263.

Marx, K. and Engels, F. (2004 [1848]) 'Manifesto of the Communist Party'. In *Marx/Engels Selected Works*, Volume 1. Marxists Internet Archive. Retrieved from www.marxists.org/archive/marx/works/1848/communist-manifesto/.

Milner, A. (2009) 'Changing the Climate: The Politics of Dystopia'. *Continuum: Journal of Media & Cultural Studies* 23 (6), 827–838.

Mumford, L. (1922) *The Story of Utopias*. New York: Boni & Liveright.

Orwell, G. (1949) *Nineteen Eighty-Four*. Oxford: Clarendon Press.

Oudenampsen, M. (2016) 'In Defence of Utopia'. *Krisis Journal for Contemporary Philosophy* 2016 (1), 43–62.

Popper, K. R. (1963) *Conjectures and Refutations: The Growth of Scientific Knowledge*. London: Routledge & Kegan Paul.

Sargent, L. T. (1994) 'Three Faces of Utopianism Revisited'. *Utopian Studies* 5 (1), 1–37.

Sargent, L. T. (2010) *Utopianism: A Very Short Introduction*. New York: Oxford University Press.

Sargisson, L. (2012) *Fool's Gold? Utopianism in the Twenty-First Century*. Basingstoke: Palgrave Macmillan.

Schapiro, L. (1972) *Totalitarianism*. London: The Pall Mall Press.

Shklar, J. N. (1989) 'The Liberalism of Fear'. In N. L. Rosenbaum (ed.) *Liberalism and the Moral Life*. Cambridge, MA: Harvard University Press.

Suvin, D. (1997) 'Locus, Horizon, and Orientation: The Concept of Possible Worlds as a Key to Utopian Studies'. In J. O. Daniel and T. Moylan (eds.) *Not Yet: Reconsidering Ernst Bloch*. London and New York: Verso, 122–137.

Talmon, J. L. (1952) *The Origins of Totalitarian Democracy*. London: Mercury Books.

Thaler, M. (2018) 'Hope Abjuring Hope: On the Place of Utopia in Realist Political Theory'. *Political Theory* 46 (5), 671–697.

Wright, E. O. (2010) *Envisioning Real Utopias*. London: Verso.

Žižek, S. (2009) 'How to Begin from the Beginning'. *New Left Review* 57, 43–55.

Žižek, S. and Thompson, P. (eds.) (2013) *The Privatization of Hope: Ernst Bloch and the Future of Utopia*. Durham, NC and London: Duke University Press.

# 2 | DISRUPTIVE UTOPIANISM: OPENING THE PRESENT

*Keijo Lakkala*

Having noted the multiple meanings of utopia and its multiple functions in facilitating social criticism, this chapter turns to broadening the discussion through the notion of utopias as a form of social practice. Specifically, utopia can be understood as a social counter-practice motivated by a desire for better being. Utopia has the potential to both relativize the current society (to distance us from the existing and given social order) and to create cracks within the present and open possibilities for new forms of being and doing. Disruption of the present opens a plurality of futures. The latter kind of utopianism (disruptive utopianism) is developed in this chapter with the concept of 'the crack', as theorized by John Holloway (2010). As will be shown, Holloway's theorizations provide a useful theoretical starting point for developing the notion of 'utopian counter-logical practice'.

I will discuss functions of utopia by moving from counter-images to counter-practices, and showing their connections. Understanding utopias as counter-images of the present means understanding them as relational, dynamic and open, as opposed to absolute, static and closed. Yet the function of utopias can be developed further, as 'lived utopianism' (Sargisson and Sargent 2017) consisting of counter-practices. In this interpretation, utopias are not just theoretically constructed counter-images of the present but formulations of alternative logics of social, economic and political practices. As relational and disruptive counter-practices, utopias open up the future by creating cracks in the social cohesion of the present.

## Utopia as Critical Counter-Image of the Present

In the Introduction, different functions of utopias were discussed. The functional approach to utopias asks: What are utopias for? This question has attracted various kinds of answers (Levitas 2010): critical and facilitating. Even within relatively coherent theoretical approaches, utopias have been seen as having both negative and positive functions. For instance within Marxism, utopias have been interpreted through the negative function of preventing social change and the positive function of facilitating such change (Levitas 2010, 6). This was noted in the Introduction by reference to nineteenth-century utopian socialists as unable to facilitate social change in Marx and Engels' (2004) eyes. The same kind of approach later made Lenin critical towards utopias in general, as he wrote in 1912: 'In politics utopia is a wish that can never come true – neither now nor afterwards, a wish that is not based on social forces and is not supported by the growth and development of political class forces' (Lenin 2004). For Lenin, utopias are mere day-dreams, fantasies and inventions that can only express the historical situation, but they cannot work politically.

The function of preventing social change is a feature which can be found mainly in the abstract and static forms of utopian thought. The negative function of utopian thought can also sometimes be understood as compensatory. This is what was also described in the Introduction by reference to Lewis Mumford's (1922, 20) concept of 'utopias of escape'. Mumford's preference, 'utopia of reconstruction', recognizes the hazards and evils of the present society and aims to reconstruct a society more suitable for human beings.

> The utopia of reconstruction is what its name implies: a vision
> of a reconstituted environment which is better adapted to
> the nature and aims of the human beings who dwell within
> it than the actual one; and not merely better adapted to their
> actual nature, but better fitted to their possible developments.
> (Mumford 1922, 21)

The utopia of escape leads back to the ego of the utopian thinker, but the utopia of reconstruction orientates towards the outside world and aims to change it.

In addition to being (more or less) facilitating, utopias always serve the function of critique (Levitas 2010, 208). Constructing a utopia is in itself implicitly critical towards the present and expresses the need for social change. Why would anyone construct utopias, if there was nothing to improve in the existing society? Utopias force a comparison of the existing society with an imaginary society, creating a contrast effect which illuminates the existing injustices. Just as Manuel and Manuel (1979, 446) write, building utopia as an antithesis to reality, a kind of counterpoint, is one of the oldest devices in the utopia writer's repertory. This contrast is already visible in Thomas More's *Utopia* (1516).

It has sometimes been said that utopias 'relativize' the present (e.g. Bauman 1976, 13), meaning that the present is always only a moment in an open-ended historical process. This has two implications: (1) the present can always be imagined to be different, and (2) the present does not determine the future, but can lead to a number of different futures depending on choices made in the present. Utopias are then tools for extrapolating the possibilities of the present. Utopias do not transcend the current reality, but draw from the experience and the cravings of their own time. The utopian ideals of an era are born from the double pressure of the real needs of that era's generation and the stubborn historical realities found in their time.

By creating a counter-image, utopias are the negation of the present. Francis and Barbara Golffing (1971) elaborate this idea in their article *An Essay on Utopian Possibility*:

Each generation entertains its own image of the future, and the image is eminently historic. Even as the world has not stood still since Campanella, or Bacon, or William Morris wrote, so neither has that counterworld – no-world, no-place (Utopos) – stood still which forms its inevitable complement. Any yes-world requires a no-world to balance it. (Golffing and Golffing 1971, 34–35)

As critical counter-images, utopias are always 'yes-worlds', in addition to being 'no-worlds'. 'Yes-world' refers here to a positive depiction of a world we desire. 'No-world', on the other hand refers to the world we want to leave behind, the troublesome present the 'yes-world' aims to overcome. A utopia is then not only a critical counter-image for highlighting the problems of its own time, but also a historical and a political goal. Utopia can be understood as a political philosophy that investigates, compares and analyzes ends, means and existing historical conditions, in order to encourage transformative action. A concern with the ends of action can be found in several imagined societies. Utopian philosophy does not just construct alternative principles for our society, but also tries to imagine what society would be like, if utopias were put into practice. When exploring these principles, utopian philosophy also questions the principles that organize existing institutions. This task of utopian philosophy has been well articulated by Peter G. Stillman:

> The utopian societies (what is not) serve as new perspectives from which to investigate the ideals, undertakings, and institutions of contemporary society, encourage a critical perspective on them, inspire a thoughtful evaluation of present and alternative individual and social ideals and activities, and consider if and where change is feasible and desirable. (Stillman 2001, 11)

## Dynamic Utopias

Utopian philosophy thereby fosters critical thinking about the existing conditions of current society and encourages envisioning alternative ways of living. It crystallizes the core problems of the present and makes us ask questions regarding the collective goals of the current society. Are they worth pursuing or could some other goals be more important? The purpose of utopian philosophy is not to create static blueprints for a new society but to create critical counter-images of a society in which radically new and better principles are put into practice. The role of utopias as counter-images of the current society makes them historically conditioned and relational.

Following philosopher and literary critic Darko Suvin, utopias can be divided into two kinds: open and closed utopias. Suvin sees no theoretical (nor empirical) grounds to see utopias as always closed and static. And even if all utopian texts would have historically been static and closed, it does not follow that this would necessarily be the case in the future too. The concept of utopia is not so much onto-logical than it is epistemological. Utopia is a thought experiment. Especially literary utopias are heuristic tools for envisioning a better world (Suvin 1997, 126–128).

The concepts of open and closed (dynamic and static) utopia can be elaborated by making a distinction between utopias focused on *locus* and those focused on utopian horizon. Utopian *locus* refers to the historical situation which shapes the utopian vision. Utopian horizon, on the other hand, means the vision in itself. The utopian horizon is dependent on the utopian *locus* and its historical development and therefore the utopian vision will be different in every historical situation. This distinction can be further developed as analysis of three necessary elements for utopian thought: (a) the place of the agent who is moving, their *locus*; (b) the horizon toward which the agent is moving; and (c) the orientation, a vector that conjoins *locus* and horizon. What is essential for the horizon is that it keeps changing as the agent moves through different *loci*.

Orientation, on the other hand, can remain more or less stable even when the space where the agent moves changes. This is why it can conjoin *locus* and horizon together. In utopian texts, the orientation of the agent is always towards a better mode of being. In utopian texts, orientation towards a better mode of being is expressed by creating possible analogical worlds to the empirical world, counter-images of the present (Suvin 1997, 130–133).

Utopia as a method for creating counter-images is always dynamic because the counter-images are always grounded in their particular historical *locus*, which they reflect negatively. The same is true also for the facilitating function of utopias. When utopias facilitate social change they do not try to create the best of all possible worlds, but to transcend the current society. In both functions, utopias are grounded in the historical situation and thereby necessarily *in relation* to their own time. They are not absolute goals for mankind but historical counter-images of the present. The point of utopias as counter-images is not to express ideas of perfection and absolute harmony but to enable distance from the existing world and to create intellectual space for thinking alternatives.

## Utopian Counter-Practice: the Crack

Utopias are not, however, just counter-*images*. The critical distance that utopias offer to the existing society can be understood on a very concrete level of social counter-*practices*. Both utopian counter-images and counter-practices are expressions of utopian desire for a better being but they also have certain dialectics between them. Counter-images can either implicitly or explicitly inform and motivate counter-practices. All counter-practices do have at least an implicit vision of what kind of world they want to create and at least an unreflected counter-image behind them that facilitates and motivates their inner logic. Often this counter-image can also take explicit, reflected and conceptually mediated forms. Counter-images can be translated into a more or less coherent set of principles, values and objectives the utopian counter-practice aims to achieve.

In this case, what needs to be observed is the inner structure and logic of these practices. Utopian qualities can be found in different communal, economic and cultural experiments, when one pays attention to the logic according to which they are carried out. These practices can be described as counter-logical social practices that clash with the logic of the present and potentially create cracks in the existing social cohesion. Utopian futures grow from these cracks.

The idea of 'the crack' is introduced by John Holloway in *Crack Capitalism* (2010). While Holloway mainly discusses strategies of social transformation, his ideas are useful for developing a conception of utopian practice. The usual understanding of social transformation has been spatial, and thereby social transformation has been associated with capturing or radically altering spaces, 'understood in traditional theory as states' (Holloway 2010, 236). However, Holloway argues that social transformation should be understood in 'apocalyptic terms', transforming not only space (state, town, or social center), but also time and relations within it. This for Holloway, means breaking duration, it means seeing 'each moment as distinct, as full of possibilities: The realization of these possibilities can mean driving each moment beyond its limits' (Holloway 2010, 236). The goal for revolution is to go beyond all limits, to the point of shedding the time itself and blending with eternity.

This transformation of time can be located in everyday behavior and practices of the present. It needs to be located in the movements, rhythms and paces of bodies in the present. It needs to be located in the logic of practices here and now. Holloway's theory is, it can be argued, useful for developing a new kind of disruptive utopianism that does not dictate the outcome of history but aims to open up the present. Not only are there utopian qualities in Holloway's theory (such as a strong desire for a better being), but it can be used to formulate a new kind of understanding of utopia. In this new understanding, utopia is not only a counter-image of the present, but also a counter-practice, motivated by a desire for better being. This counter-practice operates according to radically different logic when compared to the logic of the present.

Utopian counter-practices are about unsettling the everyday, they are about disrupting the present (see Garforth 2009).

Creating a crack on the surface of the present begins with refusal. It begins with abandoning the present and creating an alternative through revolutionary *praxis*. Holloway (2010) writes:

> Break. We want to break. We want to break the world as it is.
> A world of injustice, of war, of violence, of discrimination, of
> Gaza and Guantanamo. A world of billionaires and a billion
> people who live and die in hunger. A world in which humanity is
> annihilating itself, massacring non-human form of life, destroying
> the conditions of its own existence. A world ruled by money,
> ruled by capital. A world of frustration, of wasted potential. We
> want to create a different world. (Holloway 2010, 3)

This different world, according to Holloway, is created exactly through the 'method of the crack' (Holloway 2010, 6). The axiom of the method of the crack is that the world is always open for change, while only the ideology of the dominating class makes it seem closed and finished. The walls of the closed world are rapidly closing but the possibility for change always exists. It is only the question of revolutionary method that solves how these closing walls are to be torn down. Some revolutionaries aim to create a party led by a revolutionary avant-garde to 'denounce the movement of the walls', while others (Holloway included) 'run to the walls and try desperately to find cracks, or faults beneath the surface, or to create cracks by banging the walls' (Holloway 2010, 6). This of course relies on the assumption that such cracks are always there. Finding, locating and opening these cracks is just a matter of practical-theoretical activity, a matter of *praxis*. Theory is needed for understanding the nature of the closing wall, locating the weak spots of the wall, and practical activity is needed for creating and opening the cracks of the wall (Holloway 2010, 6).

### Cracks in Walls: Interstitial Strategy

Opening these cracks found in the 'walls' of the present is for Holloway 'the opening of a world that presents itself as closed' (Holloway 2010, 9). The method of the crack contains 'a dialectic of misfitting' (Holloway 2010, 9). This simply means to think from the point of view of those who do not fit, who are left outside. It even encourages this misfitting. It encourages non-identity, escaping from given identities. '"We" are not working class, "we" do not have a nationality. Our subjectivity cannot be reduced to the categories of the present. "We" are the non-identity, "we" are the force that contradicts all identification, the force that overflows is subjectivity' (Holloway 2009, 14).

'We', according to Holloway (2010, 9) is the indefinable subject that cannot be reduced to any given identity category. 'We' is something that could be described as negative universality. This is where the 'dialectic of misfitting' comes into play. 'We' refers to the people who refuse to fit in to capitalism. 'Ever more people simply do not fit in to the system, or, if we do manage to squeeze ourselves on to capital's ever tightening Procrustean bed, we do so at the cost of leaving fragments of ourselves behind, to haunt' (Holloway 2010, 9). The fragments that haven't fit to 'capital's ever tightening Procrustean bed' are the basis of the crack that could open a new world. 'We' who do not fit in are the basis of the crack. 'We want to understand the force of our misfitting, we want to know how banging our head against the wall over and over again will bring the wall crumbling down' (Holloway 2010, 9). 'We' is the subject that is not only unable, but also unwilling to fit in to the present capitalist society. It is the subject that screams 'No!' 'In the beginning was the scream' (Holloway 2003, 15). Screaming 'No!' is, according to Holloway (2010, 26), an act of dignity. It means that 'here and now, we refuse to subordinate our activity to the rule of capital: We can and will and are doing something else' (Holloway 2010, 26). This can be understood so that in Holloway's thought the new world is created through alternative logic of practice.

We can make distinctions between four different kinds of ways cracks can be created on the surface of the present. The first form of crack is a territorial one. Different kinds of territories occupied under a community living according to a radically different set of values can create territories in which the logic of capitalism has been overturned. Most obvious examples of these kinds of territorial cracks are the Zapatista communities in Chiapas, Mexico and the territories occupied by Kurdish militias in the Rojava region. The creation of the crack, however, cannot be thought of in terms of territory alone. Another way of thinking of it is through resource and action. The action that aims for example to decommodify vitally important resources is a form of cracking capitalism, although it cannot be reduced to any specific territory. In this situation, the crack is opened through the creation of 'commons' (Holloway 2010, 29).

Another form of crack is a temporal one. One of the most famous examples of this kind of crack is medieval festivals as described by Mikhail Bakhtin:

> On such days there is greater abundance in everything: food, dress, decorations. Festive greetings and good wishes are exchanged, although their ambivalence has faded. There are toasts, games, masquerades, laughter, pranks, and dances. The feast has no utilitarian connotation (as has daily rest and relaxation after working hours). On the contrary, the feast means liberation from all that is utilitarian, practical. It is temporary transfer to the utopian world. (Bakhtin 1984, 276)

The medieval festival is a form of temporal crack since it suspends the normal, everyday flow of things and creates a temporal, alternative world. Another example of temporal crack is what Hakim Bey (2003) calls 'Temporarily Autonomous Zones', momentary spaces of insurrection which disrupt the business as usual.

Temporality can be seen as a crucial dimension of struggle. This dimension is important when cracks are created in complex spaces. The sense of community that is, according to Holloway (2010, 30), needed for creating an autonomous zone does not typically exist in big cities, for example. 'Certainly there are plenty of spatial cracks in the cities: Social centers, squats, community gardens, publicly enjoyed spaces, but often our communities are formed on a temporal basis' (Holloway 2010, 30). These spatial cracks are usually only temporary and after a project is finished, the organizers go their different ways. However, although these cracks are only temporary, their 'rage' can 'create an otherness, a different way of doing or relating' (Holloway 2010, 30). One example of this kind of crack that has a strong temporal dimension was created during the 2001 crisis in Argentina:

> The argentinazo of 19/20 December 2001 in the cities of Argentina was not just a spatial crack, it was also a temporal

crack, a moment of rage and celebration when people descended to the streets with their pots and pans to declare that they had had enough, that all the politicians should go … and that there must be a radical change. A social energy was released, different ways of relating were created. This was a temporal crack in the patterns of domination. (Holloway 2010, 30)

The second example of cracks in the patterns of domination used by Holloway (2010, 31) is different sorts of disasters. Wars and natural disasters (earthquakes, tsunamis and hurricanes for example) do (strangely) have utopian potential, since they not only cause suffering but also 'a breakdown of social relations and the sudden emergence of quite different relations between people, relations of support and solidarity' (Holloway 2010, 31). Holloway (2010, 32) quotes Rebecca Solnit who suffered the consequences of Hurricane Juan in Halifax, Nova Scotia. According to Solnit, disasters such as Hurricane Juan suspend ordinary time and our roles and fates in society. The disasters cause obstacles to fall away and offer new possibilities of what one can do; who one might speak to; and, where one's life might be going. Everyday troubles and petty desires do not matter in a disaster situation. This is the hopeful side of these often horrible and devastating disasters. Disasters change our expectations about time and how things are supposed to work. 'The world is turned upside down just as surely as it is in a carnival: not just the physical but the social world as well … they open a window onto the possibility of another world and lay bare the miseries of the existing one' (Holloway 2010, 32).

At the moment it is, of course, the ongoing climate disaster that has, in a way, the potential to create cracks in the social cohesion of the present. Although climate change is an existential threat to humanity, it is also a force that can open the possibilities for social change. The climate crisis we are at the moment living in could create a need for new counter-images of a better, more ecologically stable society that could facilitate the masses moving towards a better future. There is, of course, a long tradition of eco-utopianism in the form of both utopian literature (see for example Callenbach 1990; McCutcheon 2015) and utopian practice (see for example Litfin 2014; Hong and Vicdan 2016), but so far these kinds of utopias have not facilitated large-scale social transformation.

The current situation is, however, so much worse than before that it could potentially work as a creative pressure to generate counter-images of the future. In fact, weak signs of this can be seen already. The IPCC Special Report from October 2018 – that demands 'rapid and far reaching transitions in energy, land, urban and infrastructure ... and industrial systems' (IPCC 2018) – has already fueled new social movements such as Extinction Rebellion who demand a radical change in the current system through the usage of non-violent strategies and tactics (Extinction Rebellion 2019). They describe themselves as 'an international apolitical network using non-violent direct action to persuade governments to act on the Climate and Ecological Emergency' (Extinction Rebellion 2019). Their vision of the future, however, is hardly utopian. It can be reduced to survival. Only vague utopian demands of 'regenerative culture' are proposed (Extinction Rebellion 2019). There is nothing inherently wrong about this, indeed it might be impossible to think about the future because of the urgency of the current threats. The problems we are dealing with right now do not offer a possibility or time to think about all-encompassing alternatives.

This experience of the urgent need for change can, however, later work as a catalyst for new future-oriented utopias. Even Extinction Rebellion might in time develop more detailed and concrete visions of the future through their Citizens' Assemblies which they use to find possible solutions to the current problems (Extinction Rebellion 2019). While this is a real possibility, at the moment no large-scale facilitating visions of a better world can be found in the arena of politics. And it is exactly this absence of mass-utopia why utopias and utopianism should focus mainly on the present. Utopianism in the form of utopian counter-images and in the form of utopian counter-logical practices can potentially 'expose the present', making visible different ways of thinking and different ways of being in the here and now. And if we can create new ways to think about the present and live in it, we can also imagine different futures. Utopian counter-logical practices can create cracks on the surface of social cohesion (and our experience of the society we live in) and give us the possibility to think and debate about the future again.

Counter-logical utopian social practices have the power to not only create cracks in the social cohesion of the existing society, but they also can potentially create cracks in our perception of the social world, in the way we experience and interact in the social world. Counter-logical utopian social practices can teach us to see the present society from a surprising perspective, and to see the possibilities for being otherwise. These practices do not, however, need to exist in a community separated from the existing society. They can also exist within the present in the form of a lived utopia.

Counter-logical utopian social practice is about creating new forms of practices within the present. It does not need to step outside of the present, but it can also work against the present within the present itself. It is about following a radically different logic of doing in the here and now that is motivated by a desire for better being. Examples of these kinds of counter-practices can be found in the theories of P2P (*peer to peer*) and timebanking, both of which abandon the logic of profit in favor of a logic of benefit in social practices. Both of these examples express the idea of the possibility of alternative logic for the present. They do not, however, imply the idea that a new community should be created outside of the present, rather they aim to turn the logics of our social practices into new configurations.

In their *Peer to Peer: The Commons Manifesto* (2019) Michel Bauwens, Vasilis Kostakis and Alex Pazaitis argue for the possibility of a new social logic of value production to emerge from within the present social world (Bauwens, Kostakis and Pazaitis 2019, 15). This new logic of value creation is based on what Bauwens, Kostakis and Pazaitis call the generative model of peer production in opposition to the extractive model of capitalism. The extractive model of capitalism relies on the logic of profit: Everything it does is aimed at maximizing profit through exploiting nature, human labor and human interaction, that is, social cooperation. The latter form of exploitation can be seen especially well in the form of cognitive capitalism that exploits networked social cooperation through unpaid activities that can be captured and financialized by proprietary 'network' platforms. Cognitive capitalism extracts the positive externalities created through human cooperation. For example, the logic of practice of many commercial social media platforms such as Facebook, Uber, Airbnb and Kickstarter is based on capturing the value of their members' social exchange; on gathering the data of their users' interactions, and then monetizing this data for profit. Cognitive capitalism focuses on the logic of extraction in every step it takes (Bauwens, Kostakis and Pazaitis 2019, 37).

The general logic of capitalism can be derived from this extractivist model of value creation: It is the logic of profit that motivates the whole capitalist social system. What Bauwens, Kostakis and Pazaitis suggest is a shift of logic from the extractivist capitalist logic to a new logic of social practice, to a new logic of production. This new logic of production can be called 'commons-based peer production' (CBPP). CBPP can be seen as a counter-practice to the extractivist logic of capitalism since it does not function according to the logic of profit, but rather according to the logic of benefit: Its priority is to produce use-values instead of exchange-values (Bauwens, Kostakis and Pazaitis 2019, 11). 'CBPP is socially embedded and oriented towards the creation of use-value. It does not rely on individual motives to gain from barter and trade to allocate resources; sharing freely is considered virtuous' (Bauwens, Kostakis and Pazaitis 2019, 15). CBPP can be described as a new logic of collaboration and social practice. It is a new way of collaboration between networks of people who freely organize around a common goal using shared resources. Examples of this kind of logic of practice can be found in such projects as Wikipedia and Linux which do not work according to the logic of profit, but according to the logic of benefit and use-value (Bauwens and Kostakis 2016, 163).

According to Bauwens, Kostakis and Pazaitis, this new logic of social practice has political effects. From 'a Gramscian perspective', they argue that CBPP can potentially have the power to advance alternatives to 'what is considered "normal" and legitimate' (Bauwens, Kostakis and Pazaitis 2019, 31). CBPP has the potential to create cracks in the social cohesion of the present and open up the possibility for a different future. It 'relativizes' the extractivist capitalist logic with a radically different logic of doing and creates a crack from which a different future can arise. Even if the future CBPP society does not as a whole evolve, it has still, by creating cracks on the surface of the social cohesion, served a political function in counter-hegemonic endeavors.

A similar counter-logic of social practice can be found in the idea of time-based currency or 'timebanks'. It too organizes itself around the logic of communal benefit instead of around the logic of profit. The core idea in timebanking is that everyone's work is of equal worth. For example, one hour of babysitting is equal to one hour of providing accounting services. This essential principle of timebanking can be seen standing against the main principles and premises of the current money system and capitalist markets, which value everyone's time and work in highly unequal ways: The completely parasitic activities of a stockbroker are valued much more highly than the necessary (and, unfortunately, unpaid) labor of mothers and other caregivers. The idea of timebanking is to provide an alternative logic of social practice that helps people to meet important personal and household needs in more socially satisfying, equal ways (Peltokoski et al. 2015).

The logic of timebanking is here contrasted against the logic of money and the premises of the current money system and capitalist markets. Timebanking offers a radically different logic of social practice, a radically different form of economic interaction. It presents a counter-logic that has the potential to clash with the logic of the present. In 2013 this clashing happened in Finland, where tax authorities of the state came out with new taxation guidelines. These new guidelines required taxing skilled work services of timebanks according to their market value (in euros) (Peltokoski et al. 2015). The state of Finland aimed to translate the logic of the social practice of timebanking to the logic of money and profit. The two logics were fundamentally incompatible and the new logic of social practice realized by timebanking clashed with the logic of the present.

In addition to CBPP and timebanking, there are plenty of other examples of what I consider to be utopian counter-logical social practices that clash with the present and have the possibility to create cracks on the surface of the present where a future could arise. Chris Carlsson has together with Francesca Manning presented various utopian counter-logical social practices: vacant-lot gardeners, 'outlaw' bicycling and cash-free gift economic practices (Carlsson and Manning 2010; see also Vaneigem 2012, 58–65). All of these practices can be seen as standing against the logic of the present society. It could be argued that they are all practices that live utopia in the here and now. These practices enter into a new world, into a world which is not based on abstract labor but on useful-creative doing, on a wholly new logic of doing. Utopia can be understood here as a world that already exists here and now, in the cracks, as a movement. To use more strategic language, one could describe this kind of utopianism 'interstitial' (see Wright 2010, 322–323).

## Conclusion

What is here called utopian counter-logical social practice is a form of 'historical experimentalism' (see Honneth 2015, 51–75) and it bases itself on the idea that history in itself does not have an inner teleology that will eventually bring us to utopia. Instead, utopian counter-logical social practices experiment with different logics of social interaction, economic activities and political decision making. The teleological idea about historical progress eventually bringing us to utopia implies a static goal that eventually wraps itself in the crust of closed totality. However utopian counter-logical social practices do not orient themselves towards a closed future state of being, but aim constantly to keep the present dynamic open for change. For utopian counter-logical social practices, the ideas of a closed totality and the end of history are inherently absurd and meaningless. The closure of the world and the end of history are impossible in the context of utopian counter-logical social practices. Utopianism is about openness and about offering radical alternatives, not about closure and perfection.

It is obvious, that for a large-scale social transformation historical experimentation with different logics of social practices is not enough. I am not advocating here exclusively 'folk political', small-scale experimentalism (see Srnicek and Williams 2015, 5–23). We will, eventually, need a collective, facilitating, future-oriented mass-utopia to tackle the global problems we are facing today. But in a situation where our social imagination has become privatized and our highest hopes seem to revolve around bare survival, the production of utopian counter-images and experimentation with counter-logical social practices have value as such. They help to distance us from the present and create cracks of out which the future can, eventually, grow. This 'opening' function is at the moment the primary function of utopianism. The closedness of the present calls for a plurality of social experimentation and utopian visions of the future. History is not teleological and there is more than one possible future ahead of us. This is where the need for utopian social imagination comes into play. In a situation where it is difficult to imagine alternatives for destructive, anthropocidal capitalism, we need to teach ourselves to dream and to imagine again. The function of utopian counter-logical social practices is to show the possibility of another world in everyday life.

# References

Bakhtin, M. (1984) *Rabelais and His World*. Bloomington, IN: Indiana University Press.

Bauman, Z. (1976) *Socialism: The Active Utopia*. London: George Allen & Unwin.

Bauwens, M. and Kostakis, V. (2016) 'Why Platform Co-ops Should Be Open Co-ops?' In T. Scholz and N. Schneider (eds.) *Ours to Hack and to Own: The Rise of Platform Cooperativism, a New Vision for the Future of Work and a Fairer Internet*. New York and London: OR Books, 163–167.

Bauwens, M., Kostakis, V. and Pazaitis, A. (2019) *Peer to Peer: The Commons Manifesto*. London: University of Westminster Press.

Bey, H. (2003) *T.A.Z. The Temporary Autonomous Zone, Ontological Anarchy, Poetic Terrorism*. New York: Autonomedia.

Callenbach, E. (1990) *Ecotopia: A Novel*. New York: Bantam Books.

Carlsson, C. and Manning, F. (2010) 'Nowtopia: Strategic Exodus?' *Antipode* 42 (4), 924–953.

Extinction Rebellion (2019) *About us*. https://rebellion.earth/the-truth/about-us/.

Garforth, L. (2009) 'No Intentions? Utopian Theory after the Future'. *Journal for Cultural Research* 13, 5–27.

Golffing, F. and Golffing, B. (1971) 'An Essay on Utopian Possibility'. In G. Kateb (ed.) *Utopia: The Potential and Prospect of the Human Condition*. New York: Routledge, 29–40.

Holloway, J. (2003) 'In the Beginning Was the Scream'. In W. Bonefeld (ed.) *Revolutionary Writing: Common Sense Essays in the Post-Political Politics*. New York: Autonomedia, 15–22.

Holloway, J. (2009) 'Why Adorno?' In J. Holloway, F. Matamoros and S. Tischler (eds.) *Negativity and Revolution: Adorno and Political Activism*. London: Pluto Press, 12–17.

Holloway, J. (2010) *Crack Capitalism*. London and New York: Pluto Press.

Hong, S. and Vicdan, H. (2016) 'Re-imagining the Utopian: Transformation of a Sustainable Lifestyle in Ecovillages'. *Journal of Business Research* 69 (1), 120–136.

Honneth, A. (2015) *The Idea of Socialism: Towards a Renewal*. Cambridge: Polity Press.

IPCC (2018) *Global Warming of 1.5°C: An IPCC Special Report on the Impacts of Global Warming of 1.5°C above Pre-industrial Levels and Related Global Greenhouse Gas Emission Pathways, in the Context of Strengthening the Global Response to the Threat of Climate Change, Sustainable Development, and Efforts to Eradicate Poverty*. Geneva: World Meteorological Organization.

Lenin, V. I. (2004) 'Two Utopias'. In *Lenin Collected Works*, Volume 18. Marxists Internet Archive. Retrieved from www.marxists.org/archive/lenin/works/1912/oct/00.htm.

Levitas, R. (2010) *The Concept of Utopia*. Bern: Peter Lang.

Litfin, K. T. (2014) *Ecovillages: Lessons for Sustainable Community*. Cambridge: Polity Press.

Manuel, F. E. and Manuel, F. P. (1979) *Utopian Thought in the Western World*. Oxford: Basil Blackwell.

Marx, K. and Engels, F. (2004) *Manifesto of the Communist Party*. In *Marx / Engels Selected Works*, Volume 1. Marxists Internet Archive. Retrieved from www.marxists.org/archive/

marx/works/1848/communist-manifesto/.

McCutcheon, E. (2015) 'More's *Utopia*, Callenbach's *Ecotopia* and Biosphere 2'. *Moreana* 52 (201/202), 149–170.

Mumford, L. (1922) *The Story of Utopias*. New York: Boni & Liveright.

Peltokoski, J., Toivakainen, N., Toivanen, T. and van der Wekken, R. (2015) 'Helsinki Timebank: Currency as a Commons'. In D. Bollier and S. Helfrich (eds.) *Patterns of Commoning*. Amherst, MA: The Commons Strategies Group and Off the Common Books.

Sargisson, L. and Sargent, L. T. (2017) 'Lived Utopianism: Everyday Life and Intentional Communities'. *Communal Societies* 37 (1), 1–24.

Srnicek, N. and Williams, A. (2015) *Inventing the Future: Postcapitalism and a World without Work*. London and New York: Verso.

Stillman, P. G. (2001) '"Nothing Is, but What Is Not"': Utopia as Practical Political Philosophy'. In B. Goodwin (ed.) *The Philosophy of Utopia*. London and New York: Routledge, 9–24.

Suvin, D. (1997) 'Locus, Horizon, and Orientation: The Concept of Possible Worlds as a Key to Utopian Studies'. In J. O. Daniel and T. Moylan (eds.) *Not Yet: Reconsidering Ernst Bloch*. London and New York: Verso, 122–137.

Vaneigem, R. (2012) *The Revolution of Everyday Life*. Oakland, CA: PM Press.

Wright, E. O. (2010) *Envisioning Real Utopias*. London and New York: Verso.

# 3 | THE PRIVATIZATION AND RECOLLECTIVIZATION OF HOPE

*Teppo Eskelinen, Keijo Lakkala and Miikka Pyykkönen*

As was noted in the Introduction, utopian thought can have better or worse times, but a general longing for a qualitatively better mode of existence is omnipresent in humanity. Yet this longing can manifest itself in various forms with greatly varying political implications. Why some manifestations become dominant, is then a key question for contemporary analysis. In this chapter, we analyze this issue using a rough distinction between two broad ways to channel the desire for better existence: first, actual (collective) utopias articulating ideas for structural societal transformations; and, second, private forms of self-development, dreams and 'escapes', often expressed without any transformative implications. Here we focus on the latter, what we call 'the privatization of utopia', reasons for its recent prominence and possible ways forward.

Much of recent research on utopias and social movements supports the observation of a general decrease in collective utopian struggles (e.g. Estlund 2014; Kamat 2014). Further, a large number of studies on social subjectivity discuss the individualization of societies, and also individualization of political dreams and aspirations (e.g. Bauman 2001; Beck and Beck-Gernsheim 2002, 1–30). People still have, of course, hopes and expectations concerning broader societal changes, but the lack of real alternatives to liberal democracies and capitalism, and the lack of sustained energy for long-term political action make people lay their concrete hopes on themselves rather than broader societal transformations. But manifestations of private hope, such as 'downshifting' or attempts to assume self-control through entrepreneurship, function as part of the contemporary system of governmentality, thus being false emancipation. Yet while noting the current 'utopia fatigue', our argument is not devoid of optimism: The will to a better existence always has the potential to be reoriented towards collective utopias.

The chapter begins with a general introduction to the state of utopias. We discuss ideas such as the 'end of history', along with different critiques of utopias, which demonstrate a given hostility towards radical social imagination within the contemporary mindset. Then we move on to discuss concrete mechanisms of privatization of utopias. The privatized life-narratives of entrepreneurship will be discussed in one section and the neoliberal 'lock-in of imagination' in the subsequent one. Towards the end of the chapter, we will discuss the potential role of social science in the task of the 'recollectivization' of utopias.

## The State of Utopias: Anxiety

As noted in the Introduction, utopia can be seen as an expression, omnipresent in humanity, of the desire for a better being (Levitas 2010, 9). As also noted, similar ideas include 'utopian propensity' (Manuel and Manuel 1979), 'the principle of Hope' (Bloch 1986) and 'utopian mentality' (Mannheim 1979, 173–236) as widely existing elements of human life with various manifestations. According to Vincent Geoghegan (2008, 17),

> we can speak of a utopian disposition, a utopian impulse or mentality, of which the classic utopia is but one manifestation. This impulse is grounded in the human capacity, and need, for fantasy; the perpetual conscious and unconscious rearranging of reality and one's place in it. It is the attempt to create an environment in which one is truly at ease.

This desire (or hope, or propensity) for a better being can be articulated in different ways, depending on ideological, political and cultural factors. Currently, with the position of liberalism as the triumphant ideology after the struggles and contradictions of the twentieth century, these articulations tend to be individualistic. This is partially an outcome of the tendency to see utopias in the absolutist sense, as essentially dangerous 'blueprints of society', as was discussed in the introduction. This anti-utopian sentiment, or 'anxiety of utopia' (Jameson 1991, 331), leads to thinking that 'the social or collective illusion of Utopia, or of a radically different society is flawed first and foremost because it is invested with a personal or existential illusion that is itself flawed from the outset' (Jameson 1991, 335).

Contextualized in contemporary society, the anxiety of utopia leads to a lack of means to think beyond capitalism. Mark Fisher coined the concept 'capitalist realism' to describe the tendency to see capitalism as the only possible mode of society. The 'realism' in Fisher's concept 'is analogous to the deflationary perspective of a depressive who believes that any positive state, any hope, is a dangerous illusion' (Fisher 2009, 5). Capitalist realism is then a mode of thought in which it is impossible to hope for different and better futures. Fisher (2009, 16) elaborates:

> Capitalist realism as I understand it cannot be confined to art or to the quasi-propagandistic way in which advertising functions. It is more like a pervasive *atmosphere*, conditioning not only the production of culture but also the regulation of work and education, and acting as a kind of invisible barrier constraining thought and action.

It is a cultural framework within which it is possible for us to think, and which sets the limits of imagination.

The flourishing of dystopian entertainment in our popular culture in recent years can be seen as a symptom of the anxiety of utopia described above. They are cultural expressions of the deep-rooted anxieties of our society (see also Laakso in this volume). As Fredric Jameson (2016, 54) writes: 'The recrudescence and reflowering of dystopias in our present culture suggests deeply rooted anxieties which are a good deal more fundamental than the fear-mongering widely practiced during the Cold War and expressed in various anti-totalitarian tracts such as Orwell's *1984*'. Utopianism can, according to Jameson (2016, 54–55), function as psychotherapy for these anxieties. 'We hammer away at anti-utopianism not with arguments, but with therapy' (Jameson 2016, 54). Jameson relies on the conventional psychoanalytic perspective according to which 'the best treatment for neurotics lies in indirection' (Jameson 2016, 54). The psychotherapist cannot tell their patients outright what is the matter with them. This will cause only resistance and raise the defenses. Instead she has to help them in their self-analysis to finally find the 'right solutions'. In psychotherapy too, according Jameson (2016, 54), it seems best to lead the patients down the wrong path from which self-knowledge can rise like a happy accident.

However, the 'anxiety of utopia' can be framed in a different manner too. Again, according to Jameson (2016, 54), on a lower and more general level it is about 'the existential fear of losing our individuality in some vaster collective'. So, what really is feared is a collective vision of future society, not a better state of being as such. What seems to have happened, then, is the re-framing of utopian desires. In the twentieth century, utopian desires for better being were typically framed in a collective manner. Both fascists and socialists framed their desires of better being with a social vocabulary. Today, the same desire for a better being is typically framed in an individualistic manner (Bauman 2017, 4).

### Mechanisms of the Privatization of Utopias: the Subjectification, Ethos and Life-Narrative of Entrepreneurship

The widespread anxiety of utopia has not come into existence without a reason and would not have been as dominant as it is without an atmosphere feeding its diffusion. But post-Cold War geopolitics is only one, and a plausibly superficial, level of explanation of the phenomenon. Cultural mechanisms lie beneath and function in tandem with political developments. So, the question is: What makes people in general direct their hopes through private rather than collective means?

Different kinds of avenues of hope are popular in different times. Whereas the ideas of changing societal and political structures and practices for the collective interests of certain group, such as the working class, were popular utopias at the turn of the twentieth century, currently the focus is on the 'personal'. If we look at the key messages of mainstream policies and social movements, manifestos for a better working life, and art projects for 'social change', for instance, we can see that to a great extent the utopian hopes and actions are placed onto an individual level (e.g. McGuigan 2009; 2016).

The conception of self-invention in this current form of 'utopianism' deviates greatly from how it is understood in collective struggles. In collective endeavors, self-invention means personal development for the purpose of social change, becoming a revolutionary subject. In the privatized ethos of hope, self-invention means personal – spiritual, corporeal or cognitive – change for becoming a better, stronger and healthier individual, without any explicit resonance to broader socio-structural transformations (Callow 2015).

If one wants influence on welfare within this privatized framework of hope, the solution is doing sports, starting a special diet, and going to yoga and mindfulness lessons. Cultural change begins and ends in the personal development of tolerance for and sensitivity to diversity or cultural civilization (e.g. admiring or practicing arts). Striving for political change means personal consumption choices and radical statements in social media. When the question is about economic change, the given answer is to be entrepreneurial.

It is easy and quick to aim at change through self-transformation. It brings more experiences and affects, which produce satisfaction and a feeling of mastering one's own life. When these positive emotions meet the growing demands on an individual's personal success, self-responsibility and a controlled risk-taking, a kind of tense field of teleological self-development is born, which largely crystallizes around the concept of 'entrepreneurship' and in the subject of 'entrepreneur'.

The success of the entrepreneur as the key figure of privatized hope is fundamentally based on three traditional elements of entrepreneurship discourse: profit-making, risk-taking and creativity (see e.g. Hisrich, Peters and Shepherd 2017; Schumpeter 1934). According to the conventional definition, entrepreneurship is a human quality, which intertwines the potential pain of risk-taking, the pleasure of using and developing one's own creativity, and the chance of basically unlimited, personal (economic) success. Nikolas Rose argued in the early 1990s that 'entrepreneurial subjectivity' – government of the individual through economic freedom, responsibility and necessity – is typical for 'advanced liberal governmentality' (Rose 1992). Many critical social scientists rediscovered the concept of 'entrepreneurship' again in the 2010s (e.g. Bröckling 2016; Olaison and Sørensen 2014; Pyykkönen 2014). Common for them is the claim that 'entrepreneurship' is rising vastly, if not already the hegemonic citizenship ethos in today's 'Western societies'. It is not only such in terms of business or work life ethos, but also extensively in terms of citizen-subjectivity in general, as the 'entrepreneurial spirit' or 'internal entrepreneurship'. Somewhat following Rose's Foucauldian path, these current social scientists claim that entrepreneurship is the 'subjective matrix' of current neoliberal governmentality, which places its conduct in developing and strengthening individual freedoms and economic capacities for enabling the greater success of 'free markets'. In other works, current governmentality works through freedom, and the entrepreneur is its key figure (Foucault 2008, 225–226).

This ethos requires and includes a personalized 'entrepreneurial utopia' (Sørensen 2009): One needs to understand oneself as a process of continuous spiritual, conceptual, attitudinal and behavioral construction and development. The horizon of this process – a productive, independent, happy and self-sufficient human being – is utopian in the sense that it can never, or at least it should not, be completely reached. More than anything else, the unbuilt hopes are a driving force for continuous personal development.

This process of change attaches to the individual impulses and dreams. The often repeated slogan of entrepreneurship educators and consultants that is 'have the courage to dream and the bravery to venture forth' encapsulates this side of the concept as well as promises that personal utopian dreams can come true with an appropriate mix of impulsiveness and risk-taking. It links implicitly to the 'overall common good' through getting its rationality from the neoliberal discourse of economic growth and competitiveness (Pyykkönen 2014). It reflects the characteristics of the neoliberal ideal citizenship (see e.g. Foucault 2008): the entrepreneurial individual, who successfully manages oneself without being a burden to society and whose life desires intertwine with the 'liberalization' of economy and public sector austerity (Kelly and Pike 2016). Sadly, this seems to be the most radical feature of individualized utopia formulated around entrepreneurship.

The entrepreneurial design of a privatized utopia would not work, if individuals would not generally think that the sacrifice required by entrepreneurship is in their own best interest (Olaison and Sørensen 2014; see also McRobbie 2016, 35–38). Hence, in pro-entrepreneurship discourse, the risks and potential negative sides are usually represented as natural characteristics of the phenomenon, which the entrepreneur-subjects have to internalize as part of their subject-building. Challenges and obstacles are central in the discourse: This way the subject of heroic narratives and representations of entrepreneur is very similar to the pioneer in classic Western narratives. The entrepreneur lives for the danger, risk-taking and gambling for collecting success and riches when the time is right – or loses it all. The possibility of bankruptcy is part of the game.

The tensions of risk-taking and surviving become nurtured by this individualized entrepreneurial utopian ethos constituted in the discourse used and produced by policy-makers, entrepreneurship organizations, self-help consultants and business gurus. In this narrative, the positive sides of entrepreneurship overcome not only its negative aspects, but also the ones of 'ordinary wage-labor'. Not all entrepreneurs internalize this hype, but elements of it often jump in one's face in entrepreneurs' pro-entrepreneurship stories in the media:

I can decide about my own working time. It is the best thing. I can sleep in the mornings, if I am tired and start the day later. I can have a day off in the middle of the week and I can basically do my work wherever I want ... I can develop all sides of my work and practically do whatever I want ... I can put all my brilliant ideas directly into practice upon my own decision ... Currently I never have to rush, because I do not intentionally create a rush for myself. (Ahokas 2019)

Being an entrepreneur makes personal freedom possible. It is something that an employee can rarely enjoy. It is freedom to decide what to do, how to do and where I do. And how much I do. Entrepreneurship gives me a privilege of coming to work happy and inspired by my work. (Stolt 2014)

The main attributes of the entrepreneurship discourse are personal freedom, creativity, and power over daily routines and time management: 'doing what you wanna do'. The entrepreneurship discourse persuades individuals with the idea that an entrepreneur, unlike the wage laborer, is not dependent on the relations of production. Other key phrases often repeated in the discourse are creativity and the ability to take care of oneself. Nevertheless, these universally positive-sounding features have their roots in business ideals and vocabulary, not any kind of 'laissez-faire' freedom or creativity.

A person with entrepreneurial spirit has a lot of those elements, that business world *requires*. A person with entrepreneurial spirit *is ready to take* responsibility, *is* motivated, *can* motivate himself, *has* clear aims, *is* aware of his goals, *develops* himself independently, *takes care* of his health. In other words, a person with entrepreneurial spirit *thinks about himself* and *builds career* as ME Ltd. weather he runs a private business or not. (Siefen 2010, italics added)

Personal freedom, autonomy and creativity are traditional propellants in all kinds of utopias – at least in those based on the voluntariness of people to a remarkable extent. For an individual, a utopia refers to a pleasant, rewarding and ideal state of being. It can be achieved through making certain things and choices (Jones and Ellis 2015). However, what distinguishes the collective socially driven utopias from entrepreneurial ethos is that the latter is stripped from the collective good and solidarity. Fundamentally the utopianism of privatized and commercialized 'entrepreneurial hope' functions on an individual level: Ideological entrepreneurship discourse promises a better personal future only for the 'true believers', i.e. those willing to perform with an entrepreneurial spirit, both in their work life and life in general.

Collective and individualized utopias differ in terms of their background rationalities: Whereas the rationality of collective utopia is a new kind of political and economic system, the rationality of private entrepreneurial utopia collapses into the political economy of the 'competition state'. The subject of collective utopia is a subject of structural change, but the entrepreneurial subject of private utopia is a subject of the structural status quo.

A further element of persuasion in the individualist hope narrative is that it recognizes an alternative to the dominant ethos, yet this alternative is similarly individualistic and entangled in the entrepreneurial hope narrative. This alternative is the idea of 'fleeing the rat-race'. Narratives of individuals 'reclaiming control of their lives' and choosing a simpler lifestyle abound in the contemporary social landscape, typically in the form of a businessman opting to 'downshift' and focus on meditation and other 'meaningful' things in life. While these narratives might not replicate the capitalist virtues of consumption and work ethic, they replicate the entrepreneurial virtues of personal freedom, autonomy and risk-taking. Not only are those virtues enforced, but seeing such 'escapes' as the only existing alternatives means seeing all existing choices as choices for the individual, with no systemic alternative to the existing social order being presented. The outcome of such private escapes is similar to 'the utopias of escape' (as discussed by Lakkala in the previous chapter): incapacity to see the prospect of societal transformation. The false dichotomy between entrepreneurialism and escapes as the choices in achieving a better mode of being does not then recognize any possibility of collective efforts of imagining and changing institutions.

## Mechanisms of the Privatization of Utopias: Retrotopia and Neoliberalism

The dominant private and personal life-projects which incarnate in the rise of the entrepreneurial mentality, of course take place within a broader cultural context. The privatization of hope requires not only discourses encouraging an individualistic and entrepreneurial spirit, but also a cultural mood directing people away from collective expressions of desires. The analysis then needs to be complemented with other key phenomena in the contemporary mental landscape feeding into the idea of the futility of collective dreaming. We suggest that two such phenomena stand out as most significant. The first is what can be called 'a nostalgic turn', or a temporal reorientation of political imagination. The second relates to the cultural outcomes of neoliberalism.

What is most characteristic about utopias, is a forward-looking temporal orientation; they portray what could be, rather than what has been. Indeed, the critical and motivating function of utopias requires that they describe a possible future. Yet exactly this capacity of looking forward seems to be challenged in the contemporary mental landscape. As has been noted by several authors writing about the state of utopias in the current era, contemporary cultural imagination seems to be more focused on looking backward (nostalgic), than looking forward (utopian) (Leone 2015; Grainge 1999). Perhaps the most notable social scientist to write about the issue is Zygmunt Bauman, who coined the term 'retrotopia' to describe this temporal turn (Bauman 2017). According to Bauman, in the age of retrotopia, societal hope is increasingly expressed in terms of 'returns' (Bauman 2017), these being calls to re-establish an imaginary past. As the alarmingly ascending far-right waves the flag of 'return to the nation-state', so does the social democratic left express its politics in terms of 'return to equality'.

Quite like fictive futures and no-place/good-places, these histories are imaginary. There never was an 'original' ethnically coherent community, quite like there never was an 'original' egalitarian welfare state. Hope expressed in terms of returns does not, then, show only a conservative wish; rather, it shows a temporal reorientation of hope. According to Bauman, the nostalgic approach to politics is 'felt at every level of social cohabitation' (Bauman 2017, 3). The implication of the emergence of nostalgia is that the future seems beyond our power to change. This does not mean that people would not prefer the capacity to imagine forward, if they possessed the collective skill to do that: Paradoxically, the ongoing era is characterized by a nostalgia for a time when we were not nostalgic (Boym 2001). This too reflects the performative power of a Fukuyama-style scholarship: 'the end of history' is not only a descriptive concept, but the prevalence of its use begins to influence human conduct and imagination. Yet again, this temporal reorientation should not be seen as an absence of the will for a better existence, but rather a symptom of the privatization of this will. The collective mode of the will for a better existence is not a sum of individual wants in a way market theory prefers to explain collective phenomena as aggregates of individual preferences. Rather, the aggregate of highly individualized – even if progressive – desires can form a society collectively longing for an imaginary past.

The nostalgic approach relates closely to neoliberalism as the dominant social form of organization and imagination. Neoliberalism is usually seen as an economic system, but its effects are not restricted to the economic sphere. Indeed, it should be seen as a process with multidimensional outcomes (e.g. Springer, Birch and MacLeavy 2016). The *material outcomes* of neoliberalism are indeed the first aspects: neoliberal policy; services and state-owned enterprises being privatized; social security systems slashed; and labor markets liberalized. This causes an increased sense of precarity and sharp polarization of incomes, wealth and access to services (Fiorentini 2015; Schatan 2001). Yet focusing on only the economic outcomes misses a dominant feature of neoliberalism, an attempt to reorganize the sphere of democratic politics. Indeed equally important outcomes of neoliberalism, as the immediate economic outcomes, are what theorists of political science call 'policy lock-in', or what was often described as the 'disciplinary' aspect of neoliberalism (Bruff 2012; Gill 1998; 2002). Neoliberalism not only operates on the level of immediate politics, but aims at changing the political framework. Neoliberal politics thus typically goes to the level of constitution: For example EU treaties, trade agreements, etc. are designed so that their modification is extremely difficult, if not impossible, by the means of normal democratic procedures; and even if this should take place, it would unleash a set of punitive measures. This system of politics is governed through increased expert power, which in itself also restricts the open political space, as several issues are depoliticized as merely 'rational'.

If the foregoing were the economic and political outcomes of neoliberalism, the third aspect, which we want to emphasize here, relates to the psychological outcomes of neoliberalism. Namely, we can ask, does neoliberal policy lock-in also cause a lock-in of imagination. This is a rarely made point, even though other psychological outcomes of the competition-oriented neoliberal society are well documented (e.g. Verhaeghe 2014). If social reality proves extremely difficult to change, does this difficulty lead us to stop even imagining what such a change could be like? Neoliberal politics have often been characterized as operating on a there-is-no-alternative mentality, yet the analysis has been seldom extended to seeing the there-is-no-possible-alternative-future aspect of these politics, even though this follows quite logically.

The lock-in of imagination is further fostered by descriptions of social reality, which naturalize social ontologies thereby marginalizing both co-operation and social change. Indeed, the neoliberal worldview is strongly grounded in an ontology based on the individual agent and her personal interests, needs and desires (Harvey 2007; McGuigan 2009, 176). This further leads to a conception of social reality in which these atomistic desires communicate only to form temporary agreements, not societal change. Especially neoclassical economics sees equilibria as the direction of motion in social structures (seen as mere aggregates of individual wants), meaning that changes always cause corresponding reactions to balance these changes. Following a similar ontology, political ideas are often represented in the form of 'political compasses', which means presenting politics as a matter of placing individuals on pre-defined (and apparently objective) axes forming a balanced space. Both ideas are performative in naturalizing an explanatory tendency, in which individuals are seen as ontologically separate from each other, with pre-existing wants or political ideas, merely seeking their position within a self-balancing structure. Indeed, if society is seen this way, it is difficult to see, what could be the starting point for an impetus for collective action aiming at social change.

## The Role of a Social Scientist

Above, we have discussed mechanisms which function to close the collective (utopian) imaginative space and enforce privatized forms of searching for a better mode of existence. These mechanisms, despite their appearance as self-control, are fundamentally mechanisms of current neoliberal governmentality. Subjectivities based on 'entrepreneurialism' eventually function to uphold a radically unimaginative society, in which it becomes normal to see politics as a social equilibrium balancing disconnected desires, or dots to be placed in space determined by fixed axes.

Yet if human desire for the better can be channeled differently, according to circumstances, it is also possible for it to reassume collective utopian modes of thought. But as the economic framework is unlikely to change rapidly, this places a responsibility on different parties to foster the skills of collective imagination. Social science then necessarily has the choice, whether it sees its purpose as only a producer of empirical data based on hegemonic ontologies, or as a force which can and should support attempts to channel utopian desires through collective means.

Social science always contributes to the generation of the social world through the production of knowledge. Yet while there is a good understanding of the necessarily engaged role of the social scientist, such action research often sees its task as finding solutions to a given specific social problem, such as a social issue experienced by a given group (e.g. Buettgen et al. 2012). As valuable as this is, a more seldom taken approach would be to take people's actual desires as a starting point and then ask, how could these desires be translated into social change? In this process of finding collective expressions for the desire for a better way of being, a possible approach would be, for instance, to try to locate their sense of escape or other dreams attached to entrepreneurship, and ask, what kind of broader social change would be relevant for these dreams, and work to fulfill them.

As noted, while the entrepreneurship discourse and its affective life-practices give personal hope and idealism for a better life, it seems hard to sketch any horizons of societal utopias from them. Yet this does not mean that these hopes and ideas could not become collective and socially progressive. The entrepreneurial utopia would need a wide acknowledgment of the collective interests and precarious lives of sole entrepreneurs, and some kind of collective strategy of improvement of their situation through societal changes. The current collective struggles of self-employed creative workers to organize unions and make general covenants for subscriptions, for instance, indicate that these kinds of changes are already taking place (Bodnar 2006) and that something which Erik Olin Wright (2010, 321–336) calls 'interstitial utopian transformation' is possible in the contexts of private forms of hope as well.

This strategy could be called a recollectivization of hope. This could also practically take place by studying small-scale initiatives for alternative organizations, and asking, how could these initiatives be upscaled (cf. Wright 2010, 273–373). So, the question would be: If, for instance a community-level initiative involves any new social relations, institutions, etc., what would these new institutions look like on a larger level? This approach also gives way to the more traditional role of the action researcher in supporting the organization of alternatives – this is worth emphasizing because of the intimate connection between imagination and organization, when it comes to constructing alternatives (Khasnabish and Haiven 2012; see also Touraine 1981).

This leads us back to the notion of utopianism. Utopianism must take an indirect approach, concentrating 'not on visions of future happiness, but rather on treatments of that stubborn resistance we tend to oppose to it and to all the other proposals for positive change' (Jameson 2016, 54). The main point of utopianism is to open up the social imagination. This means not to create fixed visions of future society or social blueprints but to create an indirect path to self-knowledge of the current society, to assist in the realization of current fears and anxieties and thereby grasp hopes and desires. '[E] very utopia today must be a psychotherapy of anti-utopian fears and draw them out into the light of day, where the sad passions like blinded snakes writhe and twist in the open air' (Jameson 2016, 55).

Another method could be the active production of 'counter-images'. Social science does not need to begin from observing the social reality, but it could assume, as a method, active reflections on what could be. This can then extend in two directions: First, assisting anyone in the actual process of constructing such counter-images, i.e. What has to be conceived within a society? What are the limits of the possible? How to stretch these limits? Second, reflecting and criticizing the present through the means of these counter-images. These ideas come very close to such modern classics as the work of Michel Foucault (1977; 1982) on the research on counter-memory, counter-conduct and resistance, and those of Alain Touraine (1981) on sociological intervention for social movements actors to become more aware of their social position and relations, their role in social change and potential consequences of their collective actions.

## Conclusions

While we took as a starting point the omnipresence of 'the desire for a different, better way of being' (Levitas 2010, 209), such desires can lose their capacity to inform actual utopias and become restricted to projects of self-realization. Privatized hope fails to be a tool for social change, rather it is a mechanism upholding existing conditions. When individual choices have no systematic connection with goals of social change, they remain directionless. While the immediate outcome of privatized utopias might be satisfaction, they represent political stagnation and a system of governance. This is particularly harmful when accompanied with inequality-generating policies, which enforce a longing for a better existence. When such alternatives are practically available only to the few, and private desires fail to be collectively communicated and formulated as social change, general frustration in society is bound to increase.

The forms and paths taken by the desire for the better are dependent on political, social and psychological conditions, largely shaping the context of imagination. Societal conditions can indeed be very disabling towards this kind of doing-by-imagining / imagining-by-doing. This disabling takes place through the creation of both social conditions and given subjectivities, and the interplay between the two. A political landscape showing few possibilities for change has an effect on imaginative skills. This is further exacerbated by a given individualistic and competitive spirit, leading to entrepreneurial subjectivities. The privatized framing of the desire for a different and better way of being is based on seeing such entrepreneurialism and private escapes as the available choices in the search for the better. No possibility of collective efforts of imagining and changing institutions or structures is then recognized.

The key issue is, whether people find common aspirations, shared language and ways to organize these aspirations. Imagining utopias is a *skill*, and therefore not an automatically existing element of human condition. As a skill, it needs to be fostered; conversely, it can also be lost. It is a crucial question, what are the mechanisms of upholding and developing this skill – in learning by doing (organizing), learning from other collective pursuits, pedagogics and so forth. Social science can have an explicit role in all this by studying and developing the elements of change and helping to foster imagination by rearticulating and contextualizing existing hopes.

## References

Ahokas, J. (2019) *Yrittäjyyden hyvät ja huonot puolet.* Blog text, March 8, 2019. Retrieved from www.jenniahokas.com/2019/03/yrittajyyden-hyvat-ja-huonot-puolet.html.

Bauman, Z. (2001) *The Individualized Society.* Cambridge: Polity Press.

Bauman, Z. (2017) *Retrotopia.* Cambridge: Polity Press.

Beck, U. and Beck-Gernsheim, E. (2002) *Individualization: Institutionalized Individualism and Its Social and Political Consequences.* London: Sage.

Bloch, E. (1986) *The Principle of Hope,* Volume 1. Cambridge, MA: The MIT Press.

Bodnar, C. (2006) 'Taking It to the Streets: French Cultural Worker Resistance and the Creation of a Precariat Movement'. *Canadian Journal of Communication* 31, 675–694.

Boym, S. (2001) *The Future of Nostalgia.* New York: Basic Books.

Bröckling, U. (2016) *The Entrepreneurial Self: Fabricating a New Type of Subject.* London: Sage.

Bruff, I. (2012) 'Authoritarian Neoliberalism, the Occupy Movements, and IPE'. *Journal of Critical Globalisation Studies* 1 (5), 114–116.

Buettgen, A., Richardson, J., Beckham, K., Richardson, K., Ward, M. and Riemer, M. (2012) 'We Did It Together: A Participatory Action Research Study on Poverty and Disability'. *Disability & Society* 27 (5), 603–616.

Callow, C. (2015) *Etherotopia, an Ideal State and a State of Mind: Utopian Philosophy as Literature and Practice.* Doctoral thesis. London: University of London.

Estlund, D. (2014) 'Utopophobia'. *Philosophy and Public Affairs* 42 (2), 113–134.

Fiorentini, R. (2015) 'Neoliberal Policies, Income Distribution Inequality and the Financial Crisis'. *Forum for Social Economics* 44 (2), 115–132.

Fisher, M. (2009) *Capitalist Realism: Is There No Alternative?* Winchester and Washington, DC: O Books.

Foucault, M. (1977) 'Nietzsche, Genealogy, History'. In D. F. Bouchard (ed.):

Language, Counter-memory, Practice: Selected Essays and Interviews by Michel Foucault. Ithaca, NY: Cornell University Press, 139–164.

Foucault, M. (1982) 'Subject and Power'. *Critical Inquiry* 8 (4), 777–795.

Foucault, M. (2008) *The Birth of Biopolitics*. New York: Picador and Palgrave Macmillan.

Geoghegan, V. (2008) *Utopianism and Marxism*. Bern: Peter Lang.

Gill, S. (1998) 'European Governance and New Constitutionalism: Economic Monetary Union and Alternatives to Disciplinary Neoliberalism in Europe'. *New Political Economy* 3 (1), 5–26.

Gill, S. (2002) 'Constitutionalizing Inequality and the Clash of Globalizations'. *International Studies Review* 4 (3), 47–65.

Grainge, P. (1999) 'Reclaiming Heritage: Colourization, Culture Wars and the Politics of Nostalgia'. *Cultural Studies* 13 (4), 621–638.

Harvey, D. (2007) *A Brief History of Neoliberalism*. Oxford: Oxford University Press.

Hisrich, R. D., Peters, M. P. and Shepherd, D. A. (2017) *Entrepreneurship*. 10th Edition. New York: McGraw-Hill Education.

Jameson, F. (1991) *Postmodernism: Or, the Cultural Logic of Late Capitalism*. Durham, NC: Duke University Press.

Jameson, F. (2016) 'An American Utopia'. In S. Žižek (ed.) *An American Utopia: Dual Power and the Universal Army*. London and New York: Verso, 1–96.

Jones, C. and Ellis, C. (2015) 'Introduction'. In C. Jones and C. Ellis (eds.) *The Individual and Utopia: A Multidisciplinary Study of Humanity and Perfection*. London: Routledge, 5–13.

Kamat, S. (2014) 'The Privatization of Public Interest: Theorizing NGO Discourse in a Neoliberal Era'. *Review of International Political Economy* 11 (1), 155–176.

Kelly, P. and Pike, J. (eds.) (2016) *Neo-Liberalism and Austerity: The Moral Economies of Young People's Health and Well-Being*. Basingstoke: Palgrave Macmillan.

Khasnabish, A. and Haiven, M. (2012) 'Convoking the Radical Imagination: Social Movement Research, Dialogic Methodologies, and Scholarly Vocations'. *Cultural Studies – Critical Methodologies* 12 (5), 408–421.

Leone, M. (2015) 'Longing for the Past: A Semiotic Reading of the Role of Nostalgia in Present-Day Consumption Trends'. *Social Semiotics* 25 (1), 1–15.

Levitas, R. (2010) *The Concept of Utopia*. Oxford: Peter Lang.

Mannheim, K. (1979) *Ideology and Utopia: An Introduction to the Sociology of Knowledge*. London and Henley: Routledge & Kegan Paul.

Manuel, F. E. and Manuel, F. P. (1979) *Utopian Thought in the Western World*. Oxford: Basil Blackwell.

McGuigan, J. (2009) *Cool Capitalism*. London: Pluto Press.

McGuigan, J. (2016) *Neoliberal Culture*. Basingstoke: Palgrave Macmillan.

McRobbie, A. (2016) *Be Creative: Making a Living in the New Culture Industries*. Cambridge: Polity Press.

Olaison, L. and Sørensen, B. (2014) 'The Abject of Entrepreneurship: Failure, Fiasco, Fraud'. *International Journal of Entrepreneurial Behavior & Research* 20 (2), 193–211.

Pyykkönen, M. (2014) *Ylistetty yrittäjyys*. Jyväskylä: SoPhi.

Rose, N. (1992) 'Governing the Enterprising Self'. In P. Heelas and P. Morris (eds.) *The Values of the Enterprise Culture: The Moral Debate*. London: Unwin Hyman, 141–164.

Schatan, J. (2001) 'Poverty and Inequality in Chile: Offspring of 25 Years of Neoliberalism'. *Development and Society* 30 (2), 57–77.

Schumpeter, J. (1934) *The Theory of Economic Development*. Cambridge, MA: Harvard University Press.

Siefen, H. (2010) *Yrittäjähenkisyys, avain ihan mihin vain*. Blog text, December 21, 2010. Retrieved from www.yrittajalinja.fi/blogi/2010/12/21/435.

Springer, S., Birch, K. and MacLeavy, J. (2016) *The Handbook of Neoliberalism*. London: Routledge.

Stolt, N. (2014) *Onnellinen yrittäjä*. Blog text, October 30, 2014. Retrieved from www.studioonni.com/onnellinen-yrittaja/.

Sørensen, B. M. (2009) 'The Entrepreneurial Utopia: Miss Black Rose and the Holy Communion'. In D. Hjorth and C. Stevaert (eds.) *The Politics and Aesthetics of Entrepreneurship*. Cheltenham: Edward Elgar, 202–220.

Touraine, A. (1981) *The Voice and the Eye: An Analysis of Social Movements*. Cambridge: Cambridge University Press.

Verhaeghe, P. (2014) *What about Me?: The Struggle for Identity in a Market-based Society*. Brunswick, Victoria: Scribe.

Wright, E. O. (2010) *Envisioning Real Utopias*. London and New York: Verso.

# PART II

# CASE STUDIES AND UTOPIAN METHODOLOGIES

# 4 | QUILOMBIST UTOPIAS: AN ETHNOGRAPHIC REFLECTION

*Inkeri Aula[1]*

Utopian imaginaries as tools for social transformation require a sense of hope. Scholarly knowledge can also be reoriented by looking for local utopias and utopian ideals embedded in particular environments and practices, and the hope they may offer. Anthropological research on hope has mainly focused on different vernacular conceptions of hope, and on the characteristics of religious and spiritual hopes (Kirksley and LeFevre 2015; Crapanzano 2004, 97–123). However, utopian hope can also be seen as a tool. This chapter takes under scrutiny how hope, understood as a form of knowledge orientation, relates to the effectiveness of utopia as a tool for social transformation.

In this chapter, I will present contemporary struggles of the Afro-Brazilian quilombist movement as an example of maintaining hope in the face of adversities. Inspired by anthropologist Hirokazu Miyazaki's 'method of hope', in which hope is taken as a method that unifies different forms of knowledge, this chapter argues that quilombismo, as it is manifested today, could bring a forward-looking orientation to the quilombos, and to Afro-Brazilian heritage more generally. Quilombist utopias put into practice in contemporary communities express hope for the future, instead of being tied to past-bound images.

The notion of the 'quilombo' has been reclaimed for decades by Afro-Brazilian movements and religious communities, black spiritual and cultural spaces, social projects and capoeira schools. The concept bears strong utopian connotations for contemporary social and cultural movements, and implies a centennial history of struggle for the ethno-territorial rights of thousands of rural black communities in Brazil. While the officially recognized quilombo communities struggle under severe oppression and violent attacks, the symbolic reclaiming of Afro-Brazilian cultural heritage connected with quilombist utopias has become ever more important. The quilombola movement defends Afro-descendant communal rights, and a way of life formed in a relation with the local natural environment in Brazil, where struggles over land are heated (e.g. Carter 2015).[2] Quilombist initiatives and influences can be found among diverse cultural expressions from capoeira and samba to political demonstrations, pointing to the significance of embodied practices entailing a social memory of Afro-Brazilian history.[3] However, translocal influences today differ from the Afrocentric quilombismo of the 1980s, so here the current phenomena is denominated 'contemporary quilombismo'. In contrast with the territorial quilombola movement, 'quilombismo' refers to a more loosely defined movement whose actors identify with Afro-descendants' struggles for political and cultural reasons.

Contemporary quilombismo is presented through an ethnographic case study, including participant observation and interviews, of two communities that claim a quilombist heritage: *Kilombo Tenondé* farm, and the cultural center *Quilombo Cecília*. Both have functioned around Capoeira Angola, another – and a transnationally better known – symbol of Afro-Brazilian resistance. Capoeira is a holistic folk art and sport that combines martial skills, dialogical movement and specific music, wherein the 'Angola' tradition emphasizes African ancestry, aesthetics and ritual aspects (Talmon-Chvaicer 2008, 151–173; Aula 2017). In the mentioned communities, foreign capoeira practitioners encounter quilombismo's antagonism to the historically constructed colonial order, implying a different way of relating with the world through embodied knowledge and Afro-Brazilian lore.

Contemporary quilombismo is asserted here as a form of utopian hope that recuperates the past-bound conception of quilombo for endeavors to maintain and develop sustainable heritages. This chapter builds on the insights of studies on marginalization and resistance in local black spaces (e.g. Leu 2014; Junior 2012; Poets 2016; Guerrón Montero 2017) by emphasizing how today social centers and cultural communities function as loci for emancipatory reinterpretations of Afro-Brazilian history and colonial globalization that constitute the utopian practice of quilombismo. Quilombismo can be analyzed as transgressive utopianism, an immanent utopia of here and now, or a prefigurative utopia not fixed into a predefined future (see Lakkala in this volume), but the relationship to the past and the future is most importantly defined by a decolonialist context of an Afro-Brazilian experience.

## Runaway Slaves or Diversified Local Economies?

The word 'kilombo' is presumably derived from a Kimbundu (spoken in Angola) word, meaning roughly 'a war camp in the wild' (Larkin Nascimento 2004, 875–876; Leite 2015, 1227; Funari 1996). Quilombos in the colonial Americas were, in iconic cases, maroon communities founded by Africans seeking freedom from slavery (e.g. Futemma, Chamlian Munari and Adams 2015; Schmitt, Manzoli Turatti and Pereira de Carvalho 2002). They were long seen as isolated black spaces of resistance, fixated in a colonial past. The Brazilian quilombo considered archetypal is *Quilombo dos Palmares*, which possibly had a population of as many as 20,000 in the seventeenth century.[4] It consisted of independent *mocambo* villages organized as a complementary economy, apparently with a fairly egalitarian social structure based partly on African kinship systems (Carneiro 1966; Funari and Vieira de Carvalho 2012).[5] Brazilian Africanists in the 1930s–1940s romanticized it as a black kingdom recreating Africa in Brazil (Carvalho 2007). This conflicted with earlier and still to some extent existing views, according to which the enslaved had been deprived of their agency, although this was disproved already by the very extension of the quilombo phenomenon.

More recent research has, however, shown that the historical quilombos were ethnically diverse and largely built with indigenous assistance; the indigenous population in Quilombo dos Palmares may have composed up to 40 percent of the population (Funari 2001; Funari and Vieira de Carvalho 2012). During the Dutch invasion of the Brazilian Northeast (1630–1650s), Dutchmen were added to the Palmarian diversity, consisting of, in addition to diverse indigenous and African peoples, European outlaws described in written documentation as 'Jews, witches, and Muslims' (Funari and Vieira de Carvalho 2012, 255; Carvalho 2007, 8). The processual formation of Afro-Brazilian culture, from the slave houses *senzalas* to the quilombo refuges, developed in interaction with the surrounding colonial economy (Reis and Santos Gomes 1996).

The image of a quilombo as a 'slave hideout' is slowly being altered to one of sovereign creation of African social agency on Brazilian soil. Anthropological studies have shown, that quilombos always had a diverse set of relationships with different rural and urban complexes occupied by Afro-Brazilian people, and many quilombos are effectuating the still ongoing urbanization tendency in Brazil as metropolitan areas swallow former rural communities (Marques and Gomes 2013; Junior 2012; Hoffman 2009; O'Dwyer 2002; Sansone 2003). Pressure from social movements (Marques and Gomes 2013, 142–143), connected with local land disputes of rural black communities organizing themselves into quilombola associations, resulted in the inclusion of quilombo land rights in the post-dictatorship Constitution of 1988 (Article 68, Brazil 1994). The official number of quilombo territories varies from 3000 to 6000, yet very few entail land ownership (Bowen 2014).[6] An obstacle in acquiring land rights has been the image of a quilombo as an isolated African society of runaway slaves, which is both suspect in the light of historical evidence and ignores the diverse composition of different black communities resisting the effects of colonialism and slavery. The quilombola movement has called for a broader conception of the quilombo for decades (Schmitt, Manzoli Turatti and Pereira de Carvalho 2002; Mendonça 2013). By the early 2000s, the narrow conception was widely contested by the anthropological reports required when applying for land rights (Linhares 2004; O'Dwyer 2002), a research tradition currently under severe political pressure. Brazil has guaranteed formal rights for indigenous and traditional groups under the ILO C169 – Indigenous and Tribal Peoples Convention – and in 2003, the land entitlement process was transmitted to INCRA, the state organ responsible for land reform (Wolford 2016).[7] To Article 68 on quilombos' land entitlement was added 'as "remainders of quilombo communities" are considered ethnic-racial groups according to criteria of self-definition, with their own historical trajectory, endowed with specific territorial relations, with the presumption of black ancestry related to resistance to the historical oppression suffered' (Article 2 Decree 4887/2003). Nevertheless, after the collective identification as quilombola, reclaiming land remained extremely difficult (Almeida et al. 2010). Ninety-five percent of the self-identified quilombo communities do not have their land ownership affirmed (Kröger and Lalander 2016).

Formally, the communities can acquire an official certification justified by an anthropological report, after which they can apply for entitlement to their communally used land. During 2004–2014, between 100 and 401 quilombo certifications were given annually, corresponding to still only 3–19 land title regularizations per year (SEPPIR 2013; INCRA 2019). After that, the situation drastically worsened. In 2016, Michel Temer's right-wing government replaced the obscurely removed President Rousseff. Long-term achievements of social movements were severely cut back, and in 2017 the quilombo land entitlement processes were halted altogether. The 2018 election of openly racist extremist Jair Bolsonaro as president increased the power of the military and the conservative 'ruralist' caucus, lobbying for the interests of grand-scale land owners. The fears of degradation of indigenous and quilombola rights have become reality, hand in hand with increasing environmental destruction and the growing power of paramilitary groups. Brazil's authoritarian government encourages openly hostile and violent attitudes towards indigenous and traditional minorities, the quilombos and their inhabitants (quilombola) among them. The murder rate of environmental activists in Brazil is among the highest in the world, and many of the current victims of these atrocities are quilombola activists.[8]

The notion of quilombo has acquired new meanings in different contexts (Guerrón Montero 2017; Assunção 2011). As Ilka Boaventura Leite (2015) claims, the term quilombo cannot be analyzed solely in relation to territory, race or an instrumentalist interest in property. Despite a variety of social problems with which the Brazilian quilombos struggle, they also promote a living community heritage. For Leite, the prevalence of the word 'quilombo' in Brazil indicates that more Brazilians are embracing an African ancestry (Leite 2015, 2). At the same time, the quilombos suffer from poverty, and rural black non-quilombo communities are displaced and persecuted.

*Quilombismo* as a political ideology was strongly formulated by Abdias do Nascimento (1980), a Brazilian pan-Africanist researcher and politician. He asserted that quilombos had a community structure based on 'African values' and a democratic organization. Their economy was the opposite of the colonial model: Instead of depending on the imperial matrix, they practiced a diversified agriculture providing their own livelihoods, and maintained relations of exchange with the surrounding populations. Nascimento's manifesto was a utopian formulation of a wider re-Africanization movement reclaiming the silenced African heritage in Brazilian society, also conjoined by the revitalization of the Capoeira Angola tradition (Aula 2017). Nascimento argued that Brazil was constructed by black peoples' forced labor, and the majority of black Brazilians continued living and working in inhuman conditions. The continuous repression had created a necessity to fight for survival. Historical quilombos imply naturally protected independent communities. However, different black organizations tolerated by authorities for religious, athletic or social purposes, have also promoted African continuity. This entire complex of 'Afro-Brazilian praxis' Nascimento denominated 'Quilombismo', calling for a 'reinvention of the Afro-Brazilian people' (Nascimento 1980, 152).

## The Method of Hope and Quilombist Knowledge

Only when I began to mature did I start to realize the political manifestations of my own people. I didn't need to go anywhere bigger. What my own people have said, written, done, this for me was the warrior's path. I was taken by Capoeira Angola. (Bahian quilombist activist)

This case study on contemporary quilombismo is founded on my fieldwork at Quilombo Cecília and Kilombo Tenondé while I was researching Afro-Bahian and European practitioners' ritual experiences on liberation in capoeira (e.g. Aula 2012; 2017). The material used in this chapter consists of participant observation records of three months in 2005 in Salvador, including interviews with Quilombo Cecília activists; fanzines; online material from 2005 to 2016 produced by the communities and a constant following of their social media; and a 40-minute interview conducted with the Kilombo Tenondé leader in 2017. Background material from Salvador includes 13 interviews, 250 photographs and audiovisual recordings.

Since the turn of the millennium, scholars have noted a lack of hope and political optimism in society and in critical research (Harvey 2000; Kirksley and LeFevre 2015; Miyazaki and Swedberg 2016). Hopelessness often results from a feeling of having no alternative to the neoliberal form of capitalism dominant today (Zournazi 2003, 135). Following anthropologist Hirokazu Miyazaki (2004), the redefinitions of quilombo can be interpreted through the theoretical idea of hope as a method of directing knowledge. Miyazaki does not approach 'hope' as an emotional state of positive expectations nor as a subject of analysis. Following philosophers Ernst Bloch, Walter Benjamin and Richard Rorty, Miyazaki takes hope to be a common operative in all knowledge formation and a *method* that unifies different forms of knowing (Miyazaki 2004, 5–7; 2006).

For the quilombist recognition of Afro-Brazilian ancestry and resistance to colonialism, the 'method of hope' dissolves a methodological problem of temporality, or the lack of a prospective orientation in analysis. Analytically, this temporal reorientation is an attempt to get closer to a fundamental question: What is knowledge for? The quilombist communities look to Afro-Brazilian history to provide solutions for the future. Quilombo Cecília was fighting for Afro-Brazilian consciousness against inequality, and Kilombo Tenondé struggles to create a sustainable way of life that could maintain diversity.

In his anthropological research on specific knowledge practices in Suvavou, Fiji, Miyazaki argues that if a hopeful state of consciousness is interpreted as an end result of a ritual, the analysis becomes incongruent with the prospective, open-to-future nature of hope itself (Miyazaki 2004, 7–12). The Suvavou people have long sought compensation for the loss of their ancestral land, where the city of Suva stands today. The fact that despite repeated rejections, they have continued to petition the government, expresses an enduring hope for confirmation of their self-knowledge on who they are and where they do belong. Failure only shows that the world is still in a Blochian state of 'not-yet' (Miyazaki 2004, 3, 31–68). For Miyazaki, maintaining prospective momentum entails an effort to replicate a past unfulfilled hope (2004, 139). The method of hope involves the inheritance of a past hope and its performative replication in the present. Hope as a method does not rest on an impulse to pursue analytical synchronicity – a correspondent image of the social world – but on an effort to inherit and replicate a spark of hope on another terrain (2004, 30). For Miyazaki, this replication reorients knowledge towards the future, in a different temporality than the current attraction of social science to follow an emergent world in a belated manner (2004, 139–140). Hope as a method, Miyazaki proposes, may help bridge the temporal gap between knowledge and its 'object'.

In Brazil, African influences are strongly present, and the majority of the population can claim some African ancestry. Dire social segregation, with white elites maintaining power while the poverty-stricken population is predominantly black, reveals the country's difficulty in dealing with the continuing effects of transatlantic slavery (Sansone 2003; Cerqueira and Moura 2013). Diverse black, indigenous, rural and environmental movements radically contest Brazil's modern development based on large-scale industries and uneven land distribution inherited from the country's colonial era (Junior 2012; Kröger and Lalander 2016). Reclaiming African ancestry, and striving for world-construction from epistemologically quilombist foundations, are attempts to occupy space for different knowledge on colonial history and structural racism, which would involve diverse corporal traditions and embodied methods of expression and narration.

The struggle for existence, as a quilombo, replicates the hopes of the freedom fighters during nearly four centuries of slavery (only abolished in Brazil in 1888 and followed by structural exclusion of the Afro-Brazilians; see Sansone 2003). Nascimento's 1980 quilombismo manifesto proposed a political mobilization of the Afro-American populations based on their own legacy, articulated as an Afro-Brazilian model for contemporary multiethnic and multicultural Brazil, basically a decolonialist blueprint for a perfect society. The manifesto articulates an ontological proposition: Another world based on mutual relations is proposed instead of an individualist, 'Eurocentric modernity'. Nascimento's principles include collective economy, anti-racism, anti-neocolonialism, anti-imperialism, and the regaining of lost self-esteem by the oppressed (1980, 168–170). For Elisa Larkin Nascimento, quilombismo, unlike European socialism or environmentalism, emphasizes harmonizing the relations between different ethnic and religious groups (Larkin Nascimento 2004, 875–878). Thus, 1980s quilombismo belongs to the 'pluriverse' as described by Arturo Escobar, worlds where lives are created in a relationship with surrounding life. In 'relational ontologies', the local particularity does not strive for domination elsewhere, but aims to live as one of many simultaneously possible worlds (as in

Afro-Colombian maroon communities; Escobar 2016). Following Escobar, relational worlds, where life is not organized by the culture/nature divide, pose an ontological struggle against the neoliberal globalizing project that treats the natural environments as unused resources (Escobar 2016, 16–23). Relational societal models and utopian practices are far from the comprehensive totalities of classic utopian forms (see Lakkala in this volume).

The non-deferrable quality of hope (Miyazaki 2004, 5) provides an openness towards the future, offering an escape from the dominant, past-bound conceptualizations of the quilombo – and of Afro-Brazilian traditions generally. A main feature associated with quilombos is antagonism to slavery. In contemporary quilombismo, this antagonism has not only been targeted towards structural racism, but also against unequal global markets, a neocolonialist and nature-exploiting world order, and the continuation of slavery in corporate Brazil. In an interconnected world of global threats, hope also needs to cross borders. The idea of orienting knowledge with hope can help us understand how quilombismo, and other social movements that draw inspiration from history, can continue to maintain prospective force, even in an extremely difficult political situation like today's Brazil (see Santos and Guarnieri 2016; Avelar 2017). Defined this way, hope could be seen as an affective background to the idea of utopias as proactive, dynamic and imaginative social criticism.

### Kilombo Tenondé: Capoeira Angola with Permaculture

Kilombo Tenondé recognizes the existence of new forms of oppression in modern, industrial society, and provides a retreat from the numbness it creates; allows for the re-generation of creativity, thought, and purpose, and a chance to rebuild community structures that have eroded.[9]

Kilombo Tenondé is a rural center for permacultural education and Capoeira Angola, located near the town of Valença, Bahia. It strives to unite people working for sustainable lifestyles, and informs them about Afro-Brazilian heritage in close interaction with the local environment. It is founded and led by the charismatic Mestre Cobra Mansa, an internationally renowned Capoeira Angola teacher in Fundação Internacional de Capoeira Angola (FICA), a capoeira school with filial groups around the world.

Kilombo Tenondé is set amidst a conserved forest and by a river with an old cottage house and new meeting spaces which are surrounded by a forest garden. International course participants learn permaculture, agroforestry and bioconstruction together with Afro-Brazilian songs, rhythms and movements imbued with transgenerational meaning. Permaculture, as in 'permanent culture' is applied worldwide as a holistic land design and community model promoting ecologically and socially sustainable life beyond economic sustainability (Holmgren 2004). Cultivation follows natural flows for the benefit of gardens and buildings; it involves creating efficient production that requires less work and no toxic substances.

Diverse organic products are crafted in the community. Overall, a balanced relationship with the natural surroundings is important in the Afro-descendant and Afro-indigenous ecological thinking in Brazil, found in *candomblé* ritual practice (e.g. Voeks 1997), and applied in the biodiverse production model of Nascimento's quilombismo (Larkin Nascimento 2004, 875–878). In the quilombola land rights movement, land is the basis of community, local economy and ethnic identity (Almeida et al. 2010). As Futemma, Chamlian Munari and Adams remark (2015), if the quilombola communities had full economic and political autonomy, they could self-govern as in the past and reconcile the goals of local development and environmental conservation.

In Brazil, quilombos are usually depicted as lacking development (Poets 2016). Due to this stereotypical backward image the quilombos share with rural indigenous and other traditional minorities, promoting quilombo heritage as a source of hope for modern society is demanding. Following Mestre Cobra Mansa's envisioning, Kilombo Tenondé is strongly focused on the mutual learning of capoeira practitioners, university students and academics, permaculture activists and locals, the latter primarily through school visits in the community. In 2016, besides the annual Permangola and Permangolinha events that combine capoeira workshops with permacultural education, courses were organized on African natural philosophy in cooperation with two universities; on Afro-Brazilian cuisine; and on diverse ways of utilizing natural materials. The Kilombo has slowly earned the respect of the local community, practicing exchange with nearby farmers, who install vending stalls at Kilombo events. People from the region's quilombola communities have attended, taking part in discussions and landrace plant exchange. This way local knowledge on living with natural surroundings can be shared.

In Kilombo Tenondé, aspiring to raise Afro-Brazilian conscious-ness seems to follow largely from Mestre Cobra Mansa's personal efforts, convincing character and networks. Accordingly, respect for the knowledge of elder cultural specialists is openly acknowledged and demanded. The expressed principles of Kilombo Tenondé include 'ancestrality, related in African-rooted and indigenous cultures to encountering nature as sacred'. There is a recognizable resonance with alternative ecological lifestyles and post-modern spirituality, but with an Afro-Brazilian emphasis. Capoeira, as an athletic bod-ily practice with a ritual flow, is imbued with manifold references to black Atlantic history, evoking mythical ancestors and events as an integral part of practitioners' lives (see Aula 2017). Connecting with ancestral temporality through corporal training is reoriented into the transformative objectives of constructing a better future: a replication of quilombist ideals. In Kilombo Tenondé, historical agroecology transformed into a permacultural practice is claimed as a quilombist continuity. Diversified horticulture, the use of local natural materials, and seasonal planning resonate not only with tra-ditional local subsistence economies of many historical quilombos, but also with an idealized model of Palmares (Nascimento 1980). This can be conceived of as a reorientation of ecological knowledge founded on the quilombo heritage but in a transnational context.

Learning natural construction and other skills functions around transnational capoeira tourism; whereas the responsibility for farm-ing is placed on a local steward and his family. Some practices of the Kilombo events are widespread in global ecovillage and alterna-tive networks, such as the sweat lodge, yoga, meditation, gatherings in circles, collecting wild horta and sharing handicraft skills. These practices are derived from traditional cultures, but practiced in a transnational alternative context, immersed in new forms of spir-ituality (Heelas 2008). For the capoeira practitioners in Tenondé, quilombist utopianism may appear as a kind of tropical paradise where, through capoeira philosophy, physical training and local self-sustainability, the world of Afro-Brazilian tradition is felt and embodied. This raises obvious questions regarding inequality of access and foreign, white and middle-class privilege. It is unclear whether communal learning beyond the logic of a profit-driven economy can help participants connect with Afro-Brazilian social issues. For foreign participants, the experience may well fall into the frame of utopian escapism as a temporary utopian community experience, as in Hakim Bey's Temporary Autonomous Zones (cited by Lakkala in this volume).

The kilombo experience may, however, have a longer-lasting effect as social criticism. At the very least, in Kilombo Tenondé capoeira practitioners, foreign and Brazilian alike, become acquainted with a different way of relating to the world beyond standard ecotourism through rhythm, movement and social narratives of resistance. This also can be asserted as 'hope', a hope for a world that offers the possibility of alternative futures that derive from local heritage. With the widespread networks of both capoeira, connected transnationally with Afro-Brazilian cultural forms and academic research, and permaculture, connected with ecological and alternative sites, the communal learning at Kilombo Tenondé spreads in a transnational web of connections. This is one face of contemporary quilombist transnationalism. The bodily training fortifies communitarian sentiments, that perhaps can lead to a recollectivization of the utopian imagination of the participants (see Eskelinen, Lakkala and Pyykkönen in this volume).

I suggest that similar novel quilombos without a direct colonial history could also be conceived of as true quilombos. This perception comes close to what Leite (2015) has named 'the post-utopian quilombo', seeking to overcome nostalgic views of the colonial past to promote instead an understanding of quilombos rooted in a radical transformation of society. It is debatable whether capoeira culture offers transformative potential beyond communal experience. The post-utopian quilombo represents 'a deconstruction of color and race as a criterion of exclusion, highlighting the quilombo as a human right', i.e. fundamental rights for Afro-Brazilian populations (Leite 2015, 1227). In a deterritorial quilombist twist we may ask: Why not also for subaltern populations elsewhere? Quilombismo as an ideology was forged in the 1980s within pan-Africanism, or calling for the union of African diasporas everywhere (inspired by such thinkers as W. E. B. DuBois, Cheikh Anta Diop and Marcus Garvey). Diversity and inclusiveness are key solutions to social problems in quilombist utopias.

## Anarchist and Afrocentric Currents in Quilombo Cecília

We plan to rescue from history all outbreaks of revolt and bring them together with the knowledge taken from our everyday experience, not in an academic, Eurocentric way, but by following the pulsations of our blood that brings with it the heritage of the mocambos, of the communes, of Sioux villages. A blood that carries the genetic memory of anarchism, capoeira, natural medicine, dance, Chiapas, punk, 1984. (Quilombola 1/1999)

For conducting ethnographic research on Afro-Brazilian heritage in transnational capoeira, I moved into the Centro Cultural Quilombo Cecília in the historical Pelourinho district of Salvador, Northeast Brazil, for three months in 2005 (Aula 2012; 2017). The social center practice resembled European anarchist meeting points: Quilombo Cecília ran a communitarian library, a video space and collection, courses and workshops, study groups, cultural activities, and offered exhibitions, publications and vegetarian lunches. In contrast with the European counterparts, the study groups dealt with black people's history, the regular activities included Capoeira Angola, and the publications dealt with African diasporas, all of which reflected the center's quilombist reference.

The textual style of the citation above is familiar from European anarchist writings and is affected by many currents, such as the Situationist International, and texts produced by the Zapatista uprising in Mexico. Another family-resemblant project, 'El Kilombo Intergaláctico' in Durham, NC, United States, demonstrates the wider resonance of similar ideas. In an interview with the Zapatista figurehead, the project is described as a social center founded by 'people of color' for educational activities: language, capoeira and computer classes, homework help, and collective decision-making in response to the neighborhood's needs, from medical consultation to housing. They cite the Zapatistas in a pro-educative saying, 'the word is our weapon'.

We were created, as a collective, in the 'todo para todos' of the Zapatistas, in the 'que se vayan todos' of the piqueteros in Argentina, in the dignity and self-respect of movements in the

United States like the Black Panthers and the Young Lords, and in the courage and commitment of all the quilombos – the indigenous, African, multi- and interracial peoples all over the world that built autonomous communities to break the relations of domination. (El Kilombo Intergaláctico 2007, vii)

In Quilombo Cecília, many of the original activities had ceased by 2005. The frustration of the quilombo activists with Eurocentric anarchism, and their subsequent interest in Afrocentric thought, introduced a dissonance that helped me expand the hermeneutic circle. Quilombo Cecília had been named after an anarchist colony founded in Brazil in the early 1900s, when many Italian anarchists immigrated to Brazil. Strongly influenced by subcultural and political movements of the 1990s, it was founded in the old town area of Salvador da Bahia in 1999. Connected to punk/hardcore and hip-hop scenes, issues like vegetarianism and natural medicine, animal rights, anti-racism, feminism and do-it-yourself culture were highlighted, and many books in the library were familiar from anarchist book fairs in Europe. In addition to study groups and courses, the Quilombo activists ran a vegetarian lunch restaurant, 'Quilombo Verde' and baked otherwise hard-to-get wholegrain bread in a bakery called, 'Conquest of Bread' after Pyotr Kropotkin's famous treatise. In Brazil, as in many other countries, the anarchist movement of the 1990s was closely connected to punk and related subcultures. Yet in Salvador relearning about the black Atlantic, Africa and Africans in diaspora was central.

Anarchism has gained increasing popularity within social movements, such as the anti-globalization movement (Graeber 2009). It is perceived as direct democracy, do-it-yourself activist culture and a non-hierarchical organization that favors consensus-based decision-making. One of the quilombo activists brought up strongly their newly found preference for an Afro-Brazilian 'natural hierarchy' based on respect for experience and age. Having lost faith in anarchism, he wanted to counter the inequalities of the stratified society in the Brazilian state of Bahia with Afro-Brazilian ideals, promoting hope deriving from the quilombo heritage. The young activists had followed the World Social Forum, but grown weary of its politics, then leaning away from the anarchist ideals of organization towards Afro-Brazilian social models in the spirit of A. Nascimento and other pan-Africanist and Afrocentric thinkers.[10] Shortly after the anarchist phase in Quilombo Cecília, its name was changed to 'Quilombo do Passo 37' and remaining activities focused on raising awareness through Afro-Brazilian cultural and academic events, including religious representatives from faiths with an African matrix. The mission, however, had not changed: It was still defined as 'the production and dissemination of culture and knowledge – to build a more humane and just society'.[11]

The tension between transnational anarchist connectivity and African ancestral inheritance makes evident the contextual particularity of specific avenues of hope. This tension was mediated by capoeira, where Afro-Brazilian ancestry became shared in the collectively produced music and martial play. The importance of corporality in the reproduction of the quilombist endeavor was described by one of the founders of Quilombo Cecília, poet Jocélia Fonseca:

> Capoeira is discovery, cure, liberation. What we do here promotes the intellect, the head. Your whole body must go along with what you say. Capoeira relates specifically to this politics of the body, in acting out this integration, everyone expressing themselves in their own way. (Salvador, March 16, 2005, author's translation)

The urge to transform society does not require any empirical basis for hope. Rather, a reorientation of knowledge occurs in replicating past struggles, as referred to by quilombo activists, on the novel terrain of contemporary activism. In the new struggles quilombismo evokes alternate knowledge on the history of slavery and resistance.

## Quilombist Hopes

Quilombist communities, including the temporary groups and events that evoke the quilombo notion that I have called here 'contemporary quilombismo', display a minoritarian social knowledge deriving from anti-colonial heritage. In anthropological discussions, the notion of 'quilombo' has been primarily connected to demands involving recognition of an Afro-Brazilian ethnic identity, livelihood and sustainable land use (e.g. O'Dwyer 2002; Linhares 2004; Almeida et al. 2010). I have argued for a need to understand

quilombo so as to encompass also novel, self-proclaimed quilombist communities with their utopian and transformative intentions. A
broader understanding of the notion has also been a central theme
for the territorial quilombola movement. I propose 'contemporary
quilombismo' to be conceived of as a loose social movement, con-
centrated around the notion of 'quilombo', that uses quilombo
utopias as tools for social transformation. Further studies are needed
to understand their possible significance and how the viability of
their proposals is assessed by traditional quilombola communities,
or the black communities currently not recognized by definitions of
quilombo. The assertion of this chapter is one of hope: The temporal
reorientation towards the future of quilombos as broadly understood could help bring hope into a seemingly hopeless moment for
the quilombo communities.

Instead of the dominant history-oriented approach to quilombos,
the prospective character of contemporary quilombismo is understood here as a method of hope as knowledge. In the quilombist
hopes for transforming society, a temporal reorientation of Afro-
Brazilian community heritage emerges. The quilombist movement
itself *is* knowledge on quilombo histories reoriented by hope. The
concept of hope makes it possible to conceive of their struggles
as prospective of a world in becoming. Adding to anthropological research on culturally specific forms of hope, hope understood
as knowledge formation illuminates formerly hidden aspects of
quilombist knowledge. Like the Fijian hopes and Japanese capitalist
dreams analyzed by Miyazaki (2004; 2006), quilombismo displays
hope as a part of temporally prospective knowledge formation. Hope
as a method is the affective epistemological background where utopia as a tool of social transformation can be implemented.

In the two self-proclaimed 'quilombos' presented here, the
embodied practice of Capoeira Angola is indispensable. I claim
that in the entanglement of capoeira with quilombos lies a deeper
cultural relation, than mere historical symbolism. This interaction
founded on a bodily practice inspires the fostering of communitarian organizations and alternative historical imagination, which is
connected to innovative responses to contemporary problems. Both
communities are also connected with other translocal movements.
The urgent hopes of translocally networking movements of permaculture and anarchism manifest themselves in these self-proclaimed

quilombos, where rereading the past fuels knowledge and action for the future. Quilombismo entangles itself with different alternative translocal movements and its connection with the martial art capoeira brings knowledge on the quilombos in an accessible way to a wider public. While rooted in history, quilombist communities maintain an openness to innovative connections and momentum for the distribution of hope.

Both examples presented here include participants from different ethnic and social backgrounds, especially through Capoeira Angola. Inclusiveness is nothing new: Historical quilombos themselves were born out of a diversity of people from different backgrounds sharing a common distress. Recent research has effectively revealed the processual and transnational development of Afro-Brazilian culture generally (Sansone 2003; Matory 2005; Aula 2017) and the vast presence of both Brazilian indigenous peoples and diverse outcast Europeans in the commonly evoked historical case of the multifaceted Quilombo dos Palmares (Funari 2001). Contemporary quilombismo echoes both the historical quilombos' antagonism to the colonial order, and their historical transnationalism.

The ideals of quilombismo do not spring from Western traditions such as dialectical materialism or the green movement, but a communal experience rooted in Afro-Brazilian history. Yet as a flexible political-cultural movement, quilombismo is in line with occidental virtues and ideals concerning human dignity. This makes it interesting not only for Afro-descendant communities, but for any culture critic exploring sustainable futures. A possibility for a reorientation of knowledge in social research comes from the hope of replicating the quilombist experience. This entails not only a reorientation of research on quilombos' history towards their struggles for existence in the future, but also a reorientation from political hopelessness to an opening for hope for futures pointed to by utopian community practices.

### Notes

1  I'm very grateful to all the colleagues who have commented on the manuscript, especially to Eeva Houtbeckers and Leo Custódio, and to my research collaborators, foremost Hélio Souza, Jocêlia Fonseca, Fábio Mandingo and Mestre Cobra Mansa – axé.

2  Conflicts involve historically marginalized rural communities, private interests and authorities, challenged by the global corporate economy (Silva et al. 2015; Bowen 2014; Haesbaert 2016).

3  Social memory is understood here as remembering shared by a group

(Järviluoma 2009). Embodied narratives can challenge the often unilinear narratives of the heritage industry (Formenti, West and Horsdal 2014; Järviluoma 2017).

4 The total population of Palmares has not been agreed upon by historians. The capital Macaco, however, had ca. 6000 inhabitants and other villages, nine in total, were smaller. See Funari (1996).

5 Through a comparison of the disparities in the oral history of Surinamese Saramakas with colonial chronicles, Richard Price argues that as with the Saramaka, there is no certainty as to if the royal families of Palmares held any executive power, nor how citizenship was understood (Price 2003).

6 Due to land conflicts, there are no official numbers on quilombo territories: population estimates vary from 1.17 million people (Fundação Palmares) to 16 million (CONAQ). Entitled quilombo lands total 1,007,827 hectares (INCRA 2019).

7 Under the Bolsonaro government the lands processed were all transferred from INCRA to the Agricultural Ministry led by the ruralist industries.

8 According to Global Witness (2018), Brazil has been the most dangerous country to be a land or environmental defender in the last decade, with a record 57 murders in 2017. The best renowned Brazilian source, however, listed 71 killings in land conflicts in 2017 with more complete information (CPT 2018). The victims are mostly indigenous people, quilombolas, landless farmers and local leaders.

9 This section's citations are founded on an interview with Mestre Cobra Mansa (April 23, 2017, 41 min.); events reported on Kilombo Tenondé's Facebook page; the website kilombotenonde.com; and the local newspaper, *Valença Agora* (2016).

10 The indispensable influence of various pan-Africanist thinkers in quilombismo, beyond the scope of this paper, calls for further studies.

11 These observations rely on interviews with Fábio Mandingo and Hélio Sousa, Salvador, March 2005, and blog entries that ran until 2009 at www.quilombo37.blogspot.com [09/2017].

## References

Almeida, A. W. B. et al. (eds.) (2010) *Territórios quilombolas e conflitos*. Cadernos de debates 1 (2). Manaus: Projeto Nova Cartografia Social da Amazônia, UEA Edições.

Assunção, M. (2011) 'Apresentação. Dossier: Novas etnicidades no Brasil: Quilombolas e índios emergentes'. *Iberoamericana* 11 (42), 85–92.

Aula, I. (2012) 'A Capoeira Angola dos praticantes europeus e bahianos: Uma comunidade transnacional experienciada'. *Antropologia Experimental* 12, 113–134.

Aula, I. (2017) 'Translocality and Afro-Brazilian Imaginaries in Globalised Capoeira'. *Suomen Antropologi: Journal of the Finnish Anthropological Society* 42 (1), 67–90.

Avelar, I. (2017) 'A Response to Fabiano Santos and Fernando Guarnieri'. *Journal of Latin American Cultural Studies* 26 (2), 341–350.

Bowen, M. (2014) 'The Struggle for Black Land Rights in Brazil: An Insider's View on Quilombos and the Quilombo Land Movement'. In F. Demissie (ed.) *African Diaspora in Brazil: History, Culture and Politics*. London: Routledge, 133–154.

Brazil (1994) *Constitution of the Federative Republic of Brazil: Announced on October 5, 1988*. São Paulo: Saraiva.

Carneiro, E. (1966) *O Quilombo dos Palmares*. Rio de Janeiro: Civilização Brasileira.

Carter, M. (ed.) (2015) *Challenging Social Inequality: The Landless Rural Workers Movement and Agrarian Reform in Brazil*. Durham, NC and London: Duke University Press.

Carvalho, A. (2007) 'Archaeological Perspectives of Palmares: A Maroon Settlement in 17th Century Brazil'. *The African Diaspora Archaeology Newsletter* 10 (1), 1–17.

Cerqueira, D. and Moura, R. (2013) *Vidas perdidas e racismo no Brasil*. Brasilia: Instituto de Pesquisa Econômica Aplicada.

CPT (2018) *Comissão Pastoral da Terra: Biblioteca virtual, caderno conflitos, assassinatos por ano*. www. cptnacional.org.br.

Crapanzano, V. (2004) *Imaginative Horizons: An Essay in Literary-Philosophical Anthropology*. Chicago, IL: University of Chicago Press.

El Kilombo Intergaláctico (2007) *Beyond Resistance: Everything. An Interview with Subcomandante Insurgente Marcos*. Durham, NC: Paperboat Press.

Escobar, A. (2016) 'Thinking-Feeling with the Earth: Territorial Struggles and the Ontological Dimension of the Epistemologies of the South'. *Revista de Antropologia Iberoamericana* 11 (1), 11–32.

Formenti, L., West, L. and Horsdal, M. (eds.) (2014) *Embodied Narratives: Connecting Stories, Bodies, Cultures and Ecologies*. Odense: University Press of Southern Denmark.

Funari, P. (1996) 'A arquelogia de Palmares'. In J. Reis and F. dos

Santos Gomes (eds.) *Liberdade por um fio: História dos quilombos no Brasil*. São Paulo: Companhia das Letras, 26–51.

Funari, P. (2001) 'Heterogeneidade e conflito na interpretação do Quilombo dos Palmares'. *Revista de História Regional* 6 (1), 11–38.

Funari, P. and Vieira de Carvalho, A. (2012) 'Gender Relations in a Maroon Community Palmares, Brazil'. In B. Voss and C. E. Casella (eds.) *The Archaeology of Colonialism: Intimate Encounters and Sexual Effects*. Cambridge: Cambridge University Press, 252–268.

Futemma, C., Chamlian Munari, L. and Adams, C. (2015) 'The Afro-Brazilian Collective Land: Analyzing Institutional Changes in the Past Two Hundred Years'. *Latin American Research Review* 50 (4), 26–48.

Global Witness (2018) *At What Cost? Irresponsible Business and the Murder of Land and Environmental Defenders in 2017*. London: Global Witness.

Graeber, D. (2009) *Direct Action: An Ethnography*. Edinburgh and Oakland, CA: AK Press.

Guerrón Montero, C. (2017) '"To Preserve Is to Resist": Threading Black Cultural Heritage from within in Quilombo Tourism'. *Souls* 19 (1), 75–90.

Haesbaert, R. (2016) 'Limites no espaço-tempo: A retomada de um debate'. *Revista Brasileira de Geografia* 61 (1), 5–20.

Harvey, D. (2000) *Spaces of Hope*. Berkeley, CA: University of California Press.

Heelas, P. (2008) *Spiritualities of Life: New Age Romanticism and Consumptive Capitalism*. Malden, MA: Wiley and Blackwell.

Hoffman, J. F. (2009) *Legalizing Identities: Becoming Black or Indian in Brazil's Northeast*. Chapel Hill, NC: University of North Carolina Press.

Holmgren, D. (2004) *Permaculture: Principles and Pathways beyond Sustainability*. Hepburn, Victoria: Holmgren Design.

INCRA (2019) *Quilombolas*. www.incra. gov.br/content/quilombolas.

Järviluoma, H. (2009) 'Soundscape and Social Memory in Skruv'. In H. Järviluoma, M. Kytö and B. Truax (eds.) *Acoustic Environments in Change*. Tampere: TAMK, 138–153.

Järviluoma, H. (2017) 'The Art and Science of Sensory Memory Walking'. In M. Cobussen, V. Meelberg and B. Truax (eds.) *The Routledge Companion to Sounding Art*. New York: Taylor & Francis, 191–204.

Junior, D. (2012) *Territorialidade, mitos e identidades coletivas: Uma etnografia de Terra de Santa na Baixada Maranhense*. M.A. dissertation. Salvador: UFBA.

Kirksley, E. and LeFevre, T. (2015) 'Reclaiming Hope: Curated Collections'. *Cultural Anthropology*. journal.culanth.org.

Kröger, M. and Lalander, R. (2016) 'Ethno-territorial Rights and the Resource Extraction Boom in Latin America: Do Constitutions Matter?' *Third World Quarterly* 37 (4), 682–702.

Larkin Nascimento, E. (2004) 'Kilombismo, Virtual Whiteness, and the Sorcery of Color'. *Journal of Black Studies* 34 (6), 861–880.

Leite, I. (2015) 'The Brazilian Quilombo: "Race", Community and Land in Space and Time'. *The Journal of Peasant Studies* 42 (6), 1225–1240.

Leu, L. (2014) 'Deviant Geographies: Black Spaces of Cultural Expression in Early 20th-Century Rio de Janeiro'. *Latin American and Caribbean Ethnic Studies* 9 (2), 177–194.

Linhares, L. (2004) 'Kilombos of Brazil: Identity and Land Entitlement'. *Journal of Black Studies* 34 (6), 817–837.

Marques, C. and Gomes, L. (2013) A Constituição de 1988 e a ressignificação dos quilombos contemporâneos: Limites e potencialidades. *Revista Brasileira de Ciências Sociais* 28 (81), 137–153.

Matory, J. L. (2005) *Black Atlantic Religion: Tradition, Transnationalism and Matriarchy in the Afro-Brazilian Candomblé*. Princeton, NJ: Princeton University Press.

Mendonça, C. (2013) *Insurgência política e desobediência epistêmica: Movimento descolonial de indígenas e quilombolas na Serra do Arapuá / PE*. Ph.D. dissertation. Recife: Universidade Federal de Pernambuco.

Miyazaki, H. (2004) *The Method of Hope: Anthropology, Philosophy, and Fijian Knowledge*. Stanford, CA: Stanford University Press.

Miyazaki, H. (2006) 'Economy of Dreams: Hope in Global Capitalism and Its Critiques'. *Cultural Anthropology* 21 (2), 147–172.

Miyazaki, H. and Swedberg, R. (eds.) (2016) *The Economy of Hope*. Philadelphia, PA: University of Pennsylvania Press.

Nascimento, A. (1980) 'Quilombismo: An Afro-Brazilian Political Alternative'. *Journal of Black Studies* 11 (2), 141–178.

O'Dwyer, E. (2002) *Quilombos: Identidade étnica e territorialidade*. Rio de Janeiro: FGV.

Poets, D. (2016) '"This Is not a Favela": Rio de Janeiro's Urban Quilombo Sacopã and the Limits of Multiculturalism'. *Bulletin of Latin American Research*, October 12.

Price, R. (2003) 'Refiguring Palmares'. *Tipití Journal of the Society for the Anthropology of Lowland South America* 1 (2), Article 3.

Reis, J. and dos Santos Gomes, F. (eds.) (1996) *Liberdade por um fio: História dos quilombos no Brasil.* São Paulo: Companhia das Letras.

Sansone, L. (2003) *Blackness without Ethnicity: Constructing Race in Brazil.* New York: Palgrave Macmillan.

Santos, F. and Guarnieri, F. (2016) 'From Protest to Parliamentary Coup: An Overview of Brazil's Recent History'. *Journal of Latin American Cultural Studies* 25 (4), 485–494.

Schmitt, A., Manzoli Turatti, M. and Pereira de Carvalho, M. (2002) 'A atualização do conceito de quilombo: Identidade e território nas definições teóricas'. *Ambiente e Sociedade* 10, 129–136.

SEPPIR (2013) *Guia de políticas públicas para comunidades quilombolas.* Brasília: SEPPIR. www.seppir.gov.br.

Silva, T., Moura, G., Oliveira, R. and Costa, E. (2015) 'Território e identidade em comunidade quilombola no Nordeste do Brasil'. *Revista Territórios & Fronteiras* 8 (2), 310–327.

Talmon-Chvaicer, M. (2008) *The Hidden History of Capoeira: A Collision of Cultures in the Brazilian Battle Dance.* Austin, TX: University of Texas Press.

Voeks, R. (1997) *Sacred Leaves of Candomblé: African Magic, Medicine and Religion in Brazil.* Austin, TX: University of Texas Press.

Wolford, W. (2016) 'The *casa* and the *causa*: Institutional Histories and Cultural Politics in Brazilian Land Reform'. *Latin American Research Review* 51 (4), 24–42.

Zournazi, M. (2003) *Hope: New Philosophies for Changes.* New York: Routledge.

# 5 | SOCIAL DREAMING AND USES OF NARRATIVITY, TELLABILITY AND EXPERIENTIALITY IN LITERARY DYSTOPIA

*Maria Laakso*

Immediately after Donald Trump's inauguration as the president of the United States in January 2017, something quite unusual happened in the American book market. Sales of dystopian classics like Aldous Huxley's *Brave New World* (1932) or Ray Bradbury's *Fahrenheit 451* (1953) skyrocketed. Within a week, George Orwell's *Nineteen Eighty-Four* (1949) spiked from Amazon's 89th most sold book to the No. 1 best-selling book. It seems that the authoritarian overtones in Trump's political campaigning increased the sales of dystopian classics instantly, once the election results revealed that this would also be the tone of forthcoming presidential politics. The sudden rise of old dystopian classics as a reaction to the contemporary political changes suggests that literary dystopia has an ability to work as a tool for reflecting the contemporary society and imagining possible futures.

By definition, literary dystopias display terrible and unpleasant worlds, conditions and societies. Furthermore, dystopian literature is typically also seen to criticize trends and flaws in contemporary society (Baccolini and Moylan 2003, 2–5; Mohr 2005, 28). Dystopia therefore not only imagines possible future conditions, but also mirrors existing reality. This kind of duality is one of its most identifiable genre features, being also characteristic of literary utopias. Dystopias are often treated as warnings about what will happen if a given practice or trend continues, or as thought experiments on the outcomes of practices or trends. However, it is important to note that dystopian fiction seems to serve its mirroring function, also for readers who can no longer relate to the temporal moment when a dystopia has been written, as is seen in the sudden rise in popularity of classic dystopias. Huxley, Orwell and Bradbury wrote their novels as a critique of a socio-political situation very different from the contemporary one. The fact that these works of speculative fiction dating back several decades still offer tools for reflecting the current political situation is interesting proof of the effectiveness and adaptability of literary fiction in general and literary dystopia in particular.

In this chapter, the relevance of literary dystopias will be elaborated. I will discuss literary dystopia as a specific manifestation of so-called *social dreaming* – 'the dreams and nightmares that concern the ways in which groups of people arrange their lives and which usually envision a radically different society than the one in which the dreams live', as Lyman Tower Sargent (1994, 3) writes. I will argue, that especially the status of literary dystopia as narrative art makes it an effective tool for reflecting social change and the possible outcomes of existing societal tendencies. Contrary to the popular perception of literary dystopias as a sign of hopelessness, I suggest that if anything is to be drawn from the sudden popularity of old dystopias, it is the readers' will to reflect their experienced societal now from a broader perspective. The willingness to imagine the future in the form of dystopian narratives is a sign of orientation towards the future, and the need to see beyond everyday practices that often prevent seeing the possibilities of change. This mirroring function of literary utopias is enabled by the experientiality of literary narration and the use of complex temporal structures. In this discussion, I will use as a reference point one of the recently most popular dystopias: *The Handmaid's Tale* (1985) by Margaret Atwood. This feminist novel, originally published in the 1980s, was already in its time a critical and popular success, and has recently become a symbol of political resistance towards Trumpian politics and current anti-feminist tendencies.[1]

I will next discuss dystopia as a subgenre of utopian thinking and the popularity of dystopias in the twenty-first century. After that, I will move to discuss *The Handmaid's Tale* to analyze how dystopian fiction offers its readers a way to evaluate ongoing societal processes even though the temporal now of the reader may differ radically from the publishing time of the dystopia. I will argue, that this is explained by literary dystopia's narrativity and experientiality, which offer a chance to reflect the ongoing now from a wider perspective: beyond personal everyday life and societal norms/al that often hide the ongoing processes. Finally, I will approach the peculiar temporality of literary dystopia, again using Atwood's novel as an example.

I argue that all dystopias have utopian undertones, because literature as a form of communication can itself be read as utopian, as it is a way of experiencing beyond one's own inevitably concise temporal and subjective perspective.

## Literary Dystopia as a Contemporary Form of Utopian Thinking

The construction of utopian ideal societies has a long history in Western literature and philosophy (see Introduction). Yet dystopia (a neologism also derived from Greek: *dis topos* – bad place) was first used much later, presumably by John Stuart Mill in 1868 (see Introduction). This historical gap is visible also in the history of utopian literature: Literary dystopias were quite rare until the end of the nineteenth century. In the family of utopian texts, dystopia can therefore be considered to be no more than a melancholic teenager, compared to the old and wise literary utopia.

The connection between these two traditions is, however, obvious. For example, M. Keith Booker (1994, 3) defines dystopia as literature that positions itself in opposition to utopian thought and warns against the potential negative results of arrant utopianism. Definitions of dystopia are often based on a few paradigmatic texts (see Lehnen 2016, 16), and arrant utopianism is surely something the 'founding fathers' of the genre criticized. Zamyatin's *We*, Huxley's *Brave New World* or Orwell's *Nineteen Eighty-Four* all imagine utopian societies gone terribly wrong and describe these societies from the perspective of suffering individuals. Even though not all contemporary dystopias follow these classics and contemporary dystopia openly borrows from other literary traditions (see Bradford et al. 2008), this thematic connection between dystopias and utopias has shadowed utopian thinking ever since the latter part of the twentieth century. The idea that 'utopia has reached a dead end' (Ferns 1999, xi) has almost become a self-fulfilling prophecy. Among utopian fiction and literary utopia, however, the claim is true. Contemporary writers rarely turn to literary utopia as a genre to imagine a better future. Neither to imagine it in satirical or critical contrast with the present, nor as literary embodiment of some social theory, nor even simply to distract imagination into alternative paths (Kumar 2013, 100–101). Dystopia, however, continues to flourish.

The one perspective that hasn't yet been fully explored in order to discuss the popularity of dystopian fiction in contemporary culture, is what could be called *the narrative appeal* of the genre. Compared to the other strands of social dreaming, literary utopia enjoyably evokes our imagination. Kumar (2013, 101) notes, that literary utopia has been the predominant form of utopia since More, because it performs 'better than any other form what utopia mostly aims to do, namely to present a "speaking picture" of the good society, to show in concrete detail what it would be like to live in such a society, and so want us to achieve it' (2013, 101). Concreteness and liveliness are surely among the appealing features of the literary form in general. Narrative form and the fictionality of literary utopias, however, don't just function as a kind of 'sugar coating designed to ease the swallowing of an otherwise indigestible bolus of political theory' (Ferns 1999, x). Instead, narratives have certain unique powers like the ability to endow experiences, perceptions, expectations and emotions with structure and ability to be shared with others.

One widely shared approach to narratives in contemporary narrative theory is the idea of close connection between narratives and minds. Narratives represent human consciousness and just as much they are tools *for* human consciousness. One of the core concepts in this sense is *experientiality* (Fludernik 1996), meaning the representation of experiencing the anthropomorphic (mostly human) element which is connected to the qualia: What something is like. David Herman (2009, 137) writes:

> Narrative representations convey the *experience* of living through storyworlds-in-flux, highlighting the pressure of events on real or imagined consciousness affected by the occurrence at issue. Thus ... it can be argued that narrative is centrally concerned with *qualia*, a term used by philosophers of mind to refer to the sense of 'what is it like' for someone or something to have particular experience.

According to this approach, literary utopia lets us not only imagine what a different society would be like, but also to know what it would be like to live in such a society and even to experience it with the mediated experience of literary characters inhabiting the fictional world. Therefore, literary utopia also encourages strong emotional engagement of the readers.

Experientiality of a narrative is always connected to the *tellability* of a narrative, which is connected to the plot. As narrative theorist Marie-Laure Ryan (1991, 148) has noted, not all plots are equal, but instead 'some configurations of facts present an intrinsic "tellability" which precedes their textualization'. The tellability of a plot is always entwined with narrative conflict. Conflicts are necessary for narrative action because a plot cannot exist without some form of conflict (Ryan 1991, 156). This leads us back to the idea of utopian societies as stable and uniform (see Introduction). From the perspective of literary and narrative studies, the stability of narrated events around the narrated utopian society is at least just as important. Bad and oppressive society offers a better chance to narrate a conflict than a stable utopian society does. If everything within a narrated society is perfect, from the perspective of an individual, the possibility of narrating a tellable conflict is less likely than within a dystopian society (Ryan 1991, 120; Schuknecht 2019, 241).

Fredric Jameson in his *The Seeds of Time* (1994, 55–56) points out, that dystopias are narrative, and utopias are mostly non-narrative. While this view is highly contested (see Moylan 2000a, 70), conflicts in dystopian fiction are often bigger than in utopias: 'the stakes are at an all-time high: the actions of one or more individual could decide the future modal structure of the entire fictional universe' (Schuknecht 2019, 239–240). Conflict is often between one man (or woman) and the entire dystopian state: Winston Smith and Big Brother in Orwell's *Nineteen Eighty-Four*; Guy Montag and the bibliophobic culture in Bradbury's *Fahrenheit 451*; or Offred and Gilead in Atwood's *The Handmaid's Tale*. The narrative structure of literary dystopias seems to offer better possibilities to build interesting conflicts than the narrative structure of literary utopias.

This leads to a suggestion, that the contemporary commercial and critical popularity of literary dystopia as well as dystopian fiction in general (movies, computer games, television) is connected to the narrative form and to their qualities as narrative literary fiction. Rafaella Baccolini and Tom Moylan (2003, 6) discuss a certain paradox involved in finding pleasure within dystopian narratives: 'Paradoxically, dystopias reach toward … the non-narrative quality of Utopia precisely by facilitating pleasurable and provocative reading experiences derived from conflicts that develop in the discrete elements of plot and character'. Dystopian fiction offers better changes for pleasurable immersion into the fictive world and to the experiences of the characters. Through the conflict-plot and character's strong emotional experiences within the dystopian world, literary dystopias are more tellable than utopias and therefore more entertaining and approachable.

### *The Handmaid's Tale*: 'A Canary in the Coal Mine'?

*The Handmaid's Tale* portrays an imaginary near-future, in which religious right-wing fundamentalists have overthrown the government of the United States and founded the Republic of Gilead in its place. Gilead is a theocratic and totalitarian society, in which women are not allowed to work, read or handle money. In this extreme version of patriarchy, fertile women are forced into sexual slavery to bear children for infertile upper-class couples. Narration also moves back in time before the revolution through the thoughts and memories of the narrator Offred – a handmaid suppressed to work as a 'two-legged womb' in a high-ranking commander's family.

As mentioned above, *The Handmaid's Tale* has become a symbol to the anti-Trump movement. In demonstrations, angry protesters carried signs such as 'Make Margaret Atwood Fiction Again' and 'The Handmaid's Tale is NOT an Instruction Manual'. The most notable of these protests was The Women's March: A worldwide protest on January 21, 2017, against anti-feminist statements made by Donald Trump and in advocacy for women's rights. Ever since, the red dresses worn by handmaids in Atwood's novel have become an identifiable symbol of women's repression. For example, in the spring of 2017, a group of female demonstrators in Texas showed up in these dresses, to fight against repressed reproductive rights, when the Texas Senate passed two anti-abortion bills.

Witty slogans discuss interestingly the complex relationships between literature and the real world. Someone might say that Atwood's novel, as well as other literary dystopias so eagerly read today, are only fiction rather than serious societal analyses of ongoing political events. But readers are well aware of this problem of referentiality. There is no doubt that readers of dystopian fiction or demonstrators carrying Atwood-signs make clear a distinction between fictional worlds and the actual world. Still dystopian fiction seems to serve as a *tool* of understanding and modeling of the contemporary society, even though the temporal now of the author might have changed; as has happened with dystopian classics read more than half a century after their publication.

Approaching literature as a *tool* leads to participating in the never-ending controversy among literary critics about whether literature should be understood as part of the sphere of aesthetics or the sphere of politics/ideology. Yet it is worth reminding, that these two notions are not exclusive, with a forced choice of one or the other. Rather, the *use of literature* should be understood in an extended way. 'Use' is not always strategic or purposeful, manipulative or grasping; it does not have to involve the sway of instrumental rationality or a willful blindness to complex form (Felski 2008, 8). Aesthetic value is inseparable from use; and engagements with texts are extraordinarily varied, complex and often unpredictable. To suggest that the meaning of literature lies in its use means to open up for investigation a vast terrain of practices, expectations, emotions, hopes, dreams and interpretations (Felski 2008, 8).

Literary dystopia has often been treated by scholars of utopia as a tool in the sense of a warning tale for adults. For instance, Rafaella Baccolini and Tom Moylan (2003, 2) write:

> the dystopian imagination has served as a prophetic vehicle, the canary in a cage, for writers with an ethical and political concern for warning us of terrible sociopolitical tendencies that could, if continued, turn our contemporary world into the iron cages portrayed in the realm of utopia's underside.

The canary in the cage is an often heard metaphor, referring to the use of canaries in coal mines, as they die from poisonous gases before the miners, who would thus be alerted to move away from the risk. Yet the danger in reducing the works of the genre of literary dystopia to mere warnings, is that the bird itself is suffocated and we don't see the other possibly important dimensions of dystopian fiction for shared social dreaming.

Especially problematic is the belief that dystopias are predictions. The only way literary fiction can predict the future is by imaging the ongoing tendencies, already happened or inventing something not yet existing. In her essay 'What "The Handmaid's Tale" Means in the Age of Trump', Atwood (2017) denies predicting the future (because of the impossibility of such a thing), but admits putting together certain existing 'strands':

> So many different strands fed into 'The Handmaid's Tale' –
> group executions. Sumptuary laws, book burnings, The
> Lebensborn program of the SS and the child stealing of
> the Argentine generals, the history of slavery, the history of
> American polygamy ... the list is long. (Atwood 2017)

Atwood claims that she has not invented things, but rather combined existing phenomena from history and the moment of writing. In this sense, the book's warning ethos is loud and clear, and the novel gets easily interpreted in its political context. *The Handmaid's Tale* certainly criticizes some political tendencies which were prevalent in Western societies at the time of the book's writing, such as the American religio-political conservative movement of the 1980s and late 1970s.

*The Handmaid's Tale* can also be interpreted as a reaction to so-called *second wave feminism*, a feminist movement that emerged around 1960 (Tolan 2007, 145; Bowman 2014, 9). Whereas first wave feminism addressed topics like labor laws, the second wave raised issues like women's sexuality, family relations and reproductive rights – all topics that *The Handmaids Tale* also addresses. The critical handling of these topics is done in the novel by mixing temporal levels of Offred's past and contemporary experiences. In the novel's narration, Offred constantly reminisces over her life in pre-Gilead society that resembles the author's contemporary world.

This way the temporal levels of the novel juxtapose flashbacks from 1970s feminist activism with the temporal now, of Offred's narration, from within an oppressive patriarchal societal system. The novel does not, however, suggest that only the male gender or patriarchy is to blame. Instead Atwood satirically criticizes the underpinning limiting utopianism of early second wave feminism (Tolan 2007, 145). Indeed, the society imagined in *The Handmaid's Tale* has unintentionally and paradoxically met certain feminist aims (Tolan 2007, 145). For example, the dystopian society is safe for women, compared to Offred's former life:

> I remember the rules, rules that were never spelled out but that
> every woman knew: don't open your door to a stranger even

if he says he is the police ... Don't stop on the road to help a
motorist pretending to be in trouble. Keep the locks on and
keep going. If anyone whistles, don't turn to look. (Atwood 2010
[1985], 36)

Even though Offred typically sees the past as a time of freedom,
this freedom also had its downside: Women were forced to live their
lives in a hostile male-dominated environment, being constantly
afraid of being raped or abused. Also the objectifying of women
exists no more in Gilead: All pornographic magazines have been
burned in bonfires and their possession is a punishable crime. In
this sense, the society has reached a feminist goal. In Atwood's
novel this censorship motif is juxtaposed with Offred's former life
via a flashback from her youth. In this flashback, Offred remembers
attending a feminist book burning with her mother and her feminist friends. The women burn magazines and other material seen as
harmful to women's liberation. The young narrator is also forced
to throw a pornographic magazine into the flames. The difference
between these two acts of censorship seems to be one of degree
(see Tolan 2007, 151).

This juxtaposition of the former societal order, with Offred's narrative present, approaches certain societal tendencies of the 1980s.
However, it is also clear, that Atwood thematizes the problematically
unobtrusive line between utopia and anti-utopia. Like the commander in the novel famously says: 'Better never means better for
everyone ... it always means worse for some' (Atwood 2010 [1985],
224). In her novel, Atwood demonstrates that every utopia is in
danger of turning into a dystopia. As Moylan (2000b, xiii) argues,
this is typical for literary dystopia and positions the genre between
utopia and anti-utopia (meaning the absolute negation of the utopia). Literary dystopia negotiates the continuum between the two
extremes, as *The Handmaid's Tale* or other dystopian classics show.
In her essay 'Dire Cartographies: The Road to Ustopia' Atwood
(2012, 66) even coins the term 'ustopia' to show how utopia and
dystopia both contain a latent version of each other. This balancing between utopia and anti-utopia could be interpreted as one of
the important aspects of the certain *usefulness* of literary dystopia as
part of social dreaming and predicting the future. Literary dystopia
reminds us of the dangers of arrant utopianism. This is one way of

explaining the societal meaningfulness readers seem to find from older dystopias. Even though, for example, *The Handmaid's Tale*'s criticism towards second wave feminism is partly outdated in the contemporary context, the broader theme of understanding utopian thinking never becomes outdated.

## 'What I Need Is a Perspective': Dystopia against the Normal and Everyday

*The Handmaid's Tale* is narrated through Offred, giving the reader direct access to the experiences of this one character. Offred thereby becomes an emotional center of the narration. One of the most terrifying scenes and also thematically loaded citations in Atwood's novel can be found at the end of the sixth chapter, where Offred stands with another handmaid, staring at the Wall where Gilead hangs the dead bodies of executed criminals. The horrifying chapter ends with the following words: 'Ordinary, said Aunt Lydia, is what you are used to. This may not seem ordinary to you now, but after a time it will. It will become ordinary' (Atwood 2010 [1985], 45).

The words of Aunt Lydia are foregrounding, as they are not framed as the direct or conscious narration of Offred, but rather these words float in the narration as free indirect discourse. Whereas Offred's narration is filtered through her critical thoughts and ironic attitude towards the repressive societal system, Aunt Lydia's inner experience remains out of the reach of the reader. For this reason, Aunt Lydia becomes the spokeswoman of the Gileadian moral order, being a woman who has internalized it. As Aunt Lydia's words show, Gileadian society doesn't in essence aim at making people believe in doing the best possible thing, but the totalitarian governance is based on an idea of the *new normal*; making people forget what once was, and thereby also the possibility of a different social order.

The concept of normality is important in understanding societal functions of dystopian fiction. Whereas utopian thinking means the capacity to imagine a future that significantly departs from general conditions of our present, *normality* means a severe threat to this capacity. As John Friedmann (2000, 462) notes, 'utopian thinking is a way of breaking through the barriers of convention into a sphere of imagination where many things beyond our everyday become feasible'. Normality breeds uniformity. As a result, there is no longing for social alternatives, and, consequently, no alternatives appear.

In order to discuss the functions of literary dystopias in social dreaming and understanding social change, it is necessary to explore the connections between *narratives* and (everyday) life. Dystopia, after all, belongs to the sphere of narrative art and narrative is a form of human cognition, 'a basic human strategy for coming to terms with time, process, and change' (Herman 2009, 2). *Narrativization* is the process of giving a narrative form to a discourse for the purpose of facilitating a better understanding of the represented phenomena (White 1980; Fludernik 1996, 31–35). Hayden White (1980, 5–6) understands the term as a transformation of historical materials into the shape of a story (and plot). In this sense, narrativization is a process of creating coherence out of raw historical data; to translate knowing into telling. As White suggests, the 'value attached to narrativity in the representations of real events arises out of a desire to have real events display the coherence, integrity, fullness, and closure of an image of a life that is and can only be imaginary' (1980, 27).

Dystopian fiction narrativizes real events or raw historical data because, as all fiction does, dystopia also imagines and invents. Monika Fludernik (1996) differentiates between historiography and narrative fiction on the basis of *experientiality*. Prototypical historiographical accounts summarize action and event sequences, but still lack some important features of experientiality, such as motivation and emotion that are typical only for fictional narratives. In literary fiction, narrativization is situated in the dynamics between the text and the reader and goes hand in hand with experientiality. Literary dystopia offers readers character-driven experiences of what it is like (qualia) to live in a dystopian society. Therefore, literary dystopia also enables empathy towards the characters (on narrative empathy see Keen 2013; Meretoja 2018).

It should be noted that the difference between life and narrative is the certain fullness of narrative compared to everyday life, which is often familiar, self-evident, concrete, contextualized and lived (Fiske 1992, 155). In the following passage, Offred reflects the past, when she was still free to work, have money and have basic human rights (like reproductive rights):

> Nothing changes instantaneously: in a gradually heating bathtub,
> you'd be boiled to death before you knew it. There were stories
> in the newspapers, of course, corpses in ditches or the woods,
> bludgeoned to death or mutilated, interfered with, as they used to

say, but they were about other women, and the men who did such things were other men. None of them were the men we knew. The newspaper stories were like dreams to us, bad dreams dreamt by others. How awful, we would say, and they were, but they were awful without being believable. They were too melodramatic, they had a dimension that was not the dimension of our lives. We were the people who were not in the papers. We lived in the blank white spaces at the edges of print. It gave us more freedom. We lived in the gaps between the stories. (Atwood 2010 [1985], 68–69)

This quotation is a good example of Atwood's ability to discuss complex issues of societal change through metaphorical language. The gradually heating bathtub is a powerful metaphor for change: always in one way or another depicted in dystopian fiction, although often as something that has already happened in an un-narrated time gap between the author's temporal now and a fictional dystopian future. It is important to notice, that societal change is here presented as something that happens *out there*, beyond our normal and beyond our everyday. Offred does not find herself in the newspaper stories. The stories would have been signals of change, warnings, but Offred and others did not identify with the stories but kept on living 'in the gaps between the stories'. Atwood claims that change happens gradually while people are too busy living their everyday lives. Here Atwood quite openly thematizes the dangers of everyday life and the lack of a wider perspective that is hard to reach from the perspective of an individual.

What I need is perspective. The illusion of depth, created by a frame, the arrangement of shapes on a flat surface. Perspective is necessary. Otherwise there are only two dimensions. Otherwise you live with your face squashed up against a wall, everything a huge foreground, of details, close-ups, hairs, the weave of the bedsheet, the molecules of the face. Your own skin like a map, a diagram of futility, crisscrossed with tiny roads that lead nowhere. Otherwise you live in the moment. Which is not where I want to be. (Atwood 2010 [1985], 155)

'Face squashed up against a wall' could be read as an intertextual reference to the Orwellian description of a future dystopia: 'boot stamping on a human face – forever'. Narrating Offred, however, is not aware of this dystopian genre that she refers to. Offred lives in a fictional world and is confined to her limited perspective. She narrates her life in the present tense (although flashbacks and memories move her narration in time) and not from a remembering stance – a safe distance from the experienced horrible events. Still she misses the perspective or the 'illusion of depth'. Atwood here discusses the important difference between life and narrativized life. As White (1980) reminds us, the world does not present itself to perception in the form of stories 'with central subjects, proper beginnings, middles, and ends, and a coherence that permits us to see the end'. Neither do all narratives, but narratives are important tools of interpreting and evaluating. Narratives give experience shape and meaning (Phelan 2008, 167). Such a narrative coherence or sense of meaning isn't something that can be reached from the perspective of the lived and experienced everyday, no matter how much Offred yearns for it. Literary fiction, dystopian fiction included, is capable of offering such a wider perspective. This is one of the key issues about the contemporary importance of dystopian fiction for social dreaming. Fiction has the ability to transcend the everyday and the limited perspective of an individual. Whereas the individual is bound to one spatial and temporal reference point of experience, fiction can offer several reference points that can together aim for a fuller and more dimensional experience of the possible future of one's own society.

Every narrative leads to the construction of a world and invites readers to imagine one (Ryan 2001, 92–93). Yet it is noteworthy, that the prototypical literary utopias of the eighteenth and nineteenth centuries, that still today shape the genre, rarely imagine a world radically different from the one inhabited by authors and readers. This is not a fault, but rather a signal of the certain function that most literary utopias serve: to 'embarrass the world we actually have' (Eagleton 2009, 33). The point is not to travel into a completely different world, but rather reflect the existing one. Therefore, dystopian fiction inevitably creates an analogue between the reader's reality and the one represented in fiction. Of course, all narrative fiction has a referential relation to the actual world outside the narrative, since the narrative comprehension is always based on real-world cognitive frames and scripts (see Ryan 1991). Literary dystopia, however, thematizes this duality through the conflict-plot that allows readers to compare their lived experiences with the characters' experiences in the narrated dystopian reality.

Also, experientiality, in the sense that Fludernik (1996, 12–13) uses the concept, is always based on real-life experience, on our embodiedness in the world. It is important, however, to notice, that experientiality is not only a quality of a text, but rather an attribute imposed on the text by the reader who actually interprets it as a narrative. Experientiality therefore has two important dimensions that are tellability and significance for the current communicational situation (narrative point). In this sense, both the literary dystopia's tellability (conflict-plot) and the dystopia's general ethos as a warning, make it an easy genre to mediate experientiality. However, because experientiality occurs in a dynamic text–reader interaction, the readers project their own experiences onto the narrative. This is one explanation to why even old literary works seem to function as relevant mirrors for current society. Dystopian fiction offers a 'wider perspective' on societal change but at the same time leaves room for its readers' own experiences of their own temporal perspectives. Therefore, no work of fiction can completely lock in the experienced fictional world but the reader fills in the gaps.

### Writing and Reading Literary Dystopia as an 'Act of Hope'

Getting back to the recent new popularity of Atwood's novel among readers as well as protesters, 'Make Margaret Atwood fiction again' is an interesting slogan. Obviously, it parodies Trump's famous 'Make America great again' slogan but when interpreted within a context of literary dystopia, it also discusses the role of fiction and the role of narratives. Compared to Atwood's novel, it is a simplification, as the limits between fiction and reality are never as clear-cut as the slogan suggests.

*The Handmaid's Tale* takes a curious turn in the last chapter called 'Historical Notes'. In this chapter, the first-person present-tense narration by Offred radically changes, as the narration takes the form of a 'transcript of the proceedings of the Twelfth Symposium on Gilead Studies' in the year 2195. The conference report shows that narration has moved into the distant future from Offred's temporal location. In 2195, Gilead has fallen, and norms suppressing women as inferiors have been subverted. Even though this last chapter reveals very little about the society it describes, the changed norms and values can be read from a keynote lecture given by the fictional history professor Pieixoto, titled 'Problems of Authentication in Reference to The Handmaid's Tale'.

The 'Handmaid's Tale' mentioned in the lecture's headline turns out to be a manuscript, a rare historical document from the long-gone patriarchal society of Gilead. As the professor in his lecture tells, it was never a written manuscript, but a sealed box with 30 cassette tapes with hours of verbal narration by a female voice. The last chapter of the novel bears many interesting meanings regarding the thematic and rhetoric of the novel. First of all, the temporal shift to the far future reveals (or rather strengthens the already revealed) ethos of the text towards the society of Gilead. The text offers the reader an exceptionally wide variety of temporal standpoints, compared to most other dystopias, that operate with the temporal now of the author, and the temporal now of the fictional world. The immediate effect of the last chapter is to give the reader a moral stance, where Gilead appears as an almost incredible societal extreme from both temporal perspectives: the reader's now and the novel's future (see Ketterer 1989, 212).

Secondly, it is important to notice that the manuscript the professor discusses is also presumably the same as what the reader has been reading until the last chapter. This adds a metafictional level to the narration, making the narrative process of the book visible. The last chapter also leads us back to the differences between different discourse forms: here historiography and narratives. Using the ideas of Fludernik, the discourse in the last chapter is not one of a literary narrative and therefore also provides less experientiality. The last chapter distances the reader from the emotionally identifiable experiences of a character suffering under the tyranny of an oppressive system. This distancing effect gives Atwood the possibility to take a new perspective from the imagined society of Gilead and to consider the complex relationship between reality and fiction – the latter also being a core question when discussing dystopias as a form of social dreaming.

The professor in the novel implies that the subjective experience of one handmaid doesn't tell the whole truth:

> As all historians know, the past is a great darkness, and filled with echoes. Voices may reach us from it; but what they say to us is imbued with the obscurity of the matrix out of which they come; and try as we may, we cannot always decipher them precisely in the clearer light of our own day. (Atwood 2010 [1985], 326)

Even though the narrator (as the professor calls Offred) tells her story, she is mute in a sense that communication is unidirectional and the voice beyond time cannot answer any questions. However, in a sense of experientiality and narrativity, the reader always becomes involved in the meaning-making process used in literature. Even though the reader is not able to communicate with the narrator, they are able to write themselves into the text to reflect their own life and the society surrounding it.

Fiction therefore is never 'just fiction', but has consequences and echoes in the actual reality. As Atwood beautifully sums up in her essay 'What "The Handmaid's Tale" Means in the Age of Trump' (2017):

> Offred records her story as best she can; then she hides it, trusting that it may be discovered later, by someone who is free to understand it and share it. This is an act of hope: Every recorded story implies a future reader. Robinson Crusoe keeps a journal. So did Samuel Pepys, in which he chronicled the Great Fire of London. So did many who lived during the Black Death, although their accounts often stop abruptly. So did Romeo Dallaire, who chronicled both the Rwandan genocide and the world's indifference to it. So did Anne Frank, hidden in her secret annex.

Examples that Atwood mentions are part of the *literature of witness*. Witness is a powerful narrative form often utilized, when art bears witness to the history and especially to traumatic history with the rhetoric of so-called *trauma narration*, that often defies representation (Felman and Laub 1992). The fact that such representations or witnesses exist, is in itself a utopian phenomenon. Literary imagination always provides experientiality and the acts of imagining future societies show that we live in a society where such a freedom of thought is accepted and encouraged.

The restriction of reading, writing and all forms of textual knowledge is a common motif in dystopian fiction: In Atwood's novel women are not allowed to read or write. The fact that they are excluded from literary representations and literary self-expression underlines the power of Offred's witness. Totalitarian dystopias imagine societies that deny the individual experience, individual values and individual meaning-making. Offred's narrative reach through time is therefore, in Atwood's own words, *an act of hope*; as is literary dystopia in general. Individual experience counteracts totalitarianism, because in its core is the unique but still providable experience. The same idea can be stretched to literary dystopia as a genre, that always has utopian undertones: As long as we *can* and *are allowed to* imagine the worst possible future worlds and societies, we are less likely to end up living in such.

## Conclusions

Above, I have discussed Atwood's *The Handmaid's Tale* and along with it the whole genre of literary dystopia as part of the wider class of narrative fiction. I have suggested that narrativity and experientiality, typical for all narrative fiction, have two kinds of consequences for our understanding of literary dystopia. First, experientiality and narrativity are connected to the plot and therefore to the pleasure of reading. Being more tellable than utopias, dystopias offer a platform for interesting conflicts and identifiable characters with strong emotions, which explains (at least partly) why dystopian fiction has acquired the status of such a dominant literary as well as fictional genre in our time.

Second, I have discussed understanding literary dystopia as a warning tool. A typical conception is that the function of dystopias is to alert us to certain tendencies in the present. However, the recent popularity of dystopian fiction suggests that literary dystopia is not as time-bound as this idea would imply. Orwell or Huxley write from a different societal perspective and still seem to offer tools for understanding our present. It rather seems that when taking the experientiality of literary dystopia into account, the meaning of the text can be located in the dynamic reading and interpretative process between the text and the reader. This helps us to understand why dystopian fiction offers important tools for social dreaming, even though the readers' temporal distance to the moment of writing can be long. Readers narrativize texts and bring to the interpretation their own experiences, which makes any literary fiction a highly adaptive tool of mind.

## Note

1 In 2019 Atwood published the novel *Testament*, a sequel to *The Handmaid's Tale*. In this article I will, however, focus on the original novel.

## References

Atwood, M. (2010 [1985]) *The Handmaid's Tale*. London: Vintage Books.

Atwood, M. (2012) *In Other Worlds: SF and the Human Imagination*. New York: Anchor Books.

Atwood, M. (2017) 'Margaret Atwood on What "The Handmaid's Tale" Means in the Age of Trump'. *The New York Times*. www.spps.org/cms/lib/MN01910242/Centricity/Domain/842/Margaret%20Atwood%20The%20New%20York%20Times.pdf.

Baccolini, R. and Moylan, T. (2003) 'Introduction. Dystopia and Histories'. In R. Baccolini and T. Moylan (eds.) *Dark Horizons: Science Fiction and the Dystopian Imagination*. London: Routledge, 1–12.

Booker, M. K. (1994) *Dystopian Literature: A Theory and Research Guide*. Westport, CT: Greenwood.

Bowman, W. (2014) 'Women and Women: Use of Women Types as Rhetorical Techniques in Atwood's Handmaid's Tale and Tepper's Gate to Woman's Country'. *Fafnir* 1 (4), 7–26.

Bradford, C., Mallan, K., Stephens, J. and McCallum, R. (2008) 'A New World Order or a New Dark Age?' In C. Bradford, K. Mallan, J. Stephens and R. McCallum (eds.) *New World Orders in Contemporary Children's Literature: Utopian Transformations*. London: Palgrave Macmillan, 1–10.

Eagleton, T. (2009) 'Utopia and Its Opposites'. In L. Panitchs and C. Leys (eds.) *Socialist Register 2000: Necessary and Unnecessary Utopias. Social Register* Volume 36, 31–40.

Felman, S. and Laub, D. (1992) *Testimony: Crises of Witnessing in Literature, Psychoanalysis and History*. London: Routledge.

Felski, R. (2008) *Uses of Literature*. Oxford: Blackwell.

Ferns, C. (1999) *Narrating Utopia: Ideology, Gender, Form in Utopian Literature*. Liverpool: Liverpool University Press.

Fiske, J. (1992) 'Cultural Studies and the Culture of Everyday Life'. In L. Grossberg (ed.) *Cultural Studies*. London and New York: Routledge, 38–55.

Fludernik, M. (1996) *Towards a 'Natural' Narratology*. London: Routledge.

Friedmann, J. (2000) 'The Good City: In Defense of Utopian Thinking'. *International Journal of Urban and Regional Research* 24, 460–472.

Herman, D. (2009) *The Basic Elements of Narrative*. Malden, MA: Wiley-Blackwell.

Jameson, F. (1994) *The Seeds of Time*. New York: Columbia University Press.

Keen, S. (2013) 'Narrative Empathy'. In *The Living Handbook of Narratology*. www.lhn.uni-hamburg.de/node/42.html.

Ketterer, D. (1989) 'Margaret Atwood's "The Handmaid's Tale": A Contextual Dystopia'. *Science Fiction Studies* 16 (2), 209–217.

Kumar, K. (2013) 'The Future of Utopia'. In J. Bastos da Silva (ed.) *The Epistemology of Utopia: Rhetoric, Theory and Imagination*. Newcastle upon Tyne: Cambridge Scholars Publishing, 94–119.

Lehnen, C. (2016) *Defining Dystopia: A Genre between the Circle and the Hunger Games. A Functional Approach to Fiction*. Marburg: Tectum Verlag.

Meretoja, H. (2018) *The Ethics of Storytelling: A Narrative*

Hermeneutics, History and the Possible. New York: Oxford University Press.

Mohr, D. M. (2005) *Words Apart? Dualism and Transgression in Contemporary Female Dystopias.* London: McFarland & Company.

Moylan, T. (2000a) 'A Look into the Dark: On Dystopia and the Novum'. In P. Parrinder (ed.) *Learning from Other Worlds.* Durham, NC: Duke University Press, 51–71.

Moylan, T. (2000b) *Scraps of the Untainted Sky: Science Fiction, Utopia, Dystopia.* Boulder, CO: Westview Press.

Orwell, G. (1949) *Nineteen Eighty-Four.* Oxford: Clarendon Press.

Phelan, J. (2008) 'Narratives in Contest; or Another Twist in the Narrative Turn'. *PMLA* 123 (1), 166–175.

Ryan, M.-L. (1991) *Possible Worlds, Artificial Intelligence, and Narrative Theory.* Bloomington, IN: Indiana University Press.

Ryan, M.-L. (2001) *Narrative as Virtual Reality: Immersion and Interactivity in Literature and Electronic Media.* Baltimore, MD and London: The Johns Hopkins University Press.

Sargent, L. T. (1994) 'The Three Faces of Utopianism Revisited'. *Utopian Studies* 5 (1), 1–37.

Schuknecht, M. (2019) 'The Best/Worst of All Possible Worlds? Utopia, Dystopia, and Possible Worlds Theory'. In A. Bell and M.-L. Ryan (eds.) *Possible Worlds Theory and Contemporary Narratology.* Lincoln, NE and London: Nebraska University Press, 225–248.

Tolan, F. (2007) *Margaret Atwood: Feminism and Fiction.* Amsterdam and New York: Rodopi.

White, H. (1980) 'The Value of Narrativity in the Representation of Reality'. *Critical Inquiry* 7 (1), 5–27.

# 6 | UTOPIAN EDUCATION: MAY THE HOPE BE WITH YOU

## Olli-Pekka Moisio and Matti Rautiainen

Educators work in a world facing challenges such as climate change, dramatically affecting our life now and in the future. Educators construct the future world through their work, because they prepare young generations for lives in a world that does not yet exist. Additionally, they educate children to live in the current world. This active role requires a combination of futurist imagination and its implementation in everyday life. This dual role has always been part of formal education praxis. For example in Finland, twenty-first century skills have been emphasized in the documents of educational policy along with application in the curricula of basic education and secondary schools in the 2010s (National Core Curriculum for Basic Education 2016). Therefore, the relationship between education and utopia is strong both theoretically and practically.

In utopian literature it is typical to describe the ideal state more precisely than the means to reach that state. Since the earliest utopias, like Plato's description of Atlantis, utopias have included the key objectives of education, which make a utopia possible. From the nineteenth century onwards, new political ideologists, especially within the socialist movement, focused also on how the state of ideology (utopia) would be created and how education would promote the new society (see e.g. Kropotkin 1974 [1912]). In the twentieth century, when political ideologies shifted to meet the educational policies of nation-states, utopian ideas became concrete goals for educational policy and practice in the everyday life of educational institutions. Finland displays a typical example of this movement, and, for example, curricula for basic education reflected strongly concepts of hope, equality, democracy and the future.

In this chapter, we study the relationship between utopia and education from three viewpoints. First, the question of human existence, especially hope and despair, which are at the core of education and utopia. Second, Paulo Freire's pedagogical thinking, as Freire created a concept of utopian education for educators to use utopian thinking in their work. Third, we introduce a case study example on how utopia can be visible in the everyday life of educators.

## Forward Dawning

At the junction of past and future, our lives merge with fleeting moments. We wait for something to arrive. Anticipation is communication between past and future. This is beautifully illustrated in Leo Tolstoy's *Anna Karenina*. At the end of Part 7, Anna arrives at the railroad station (a constant motif in the novel):

> Walking through the crowd into the first-class waiting room, she gradually recalled all the details of her situation and the decisions among which she had been hesitating. And first hope, then despair over old hurts again began to chafe the wounds of her tormented, terribly fluttering heart. Sitting on a star-shaped sofa and waiting for the train, looking with revulsion at the people coming in and going out (they all disgusted her), she thought of how she would arrive at the station, write a note to him, and of what she would write, then of how he was now complaining to his mother (not understanding her suffering) about his situation, and how she would come into the room and what she would say to him. Then she thought of how life could still be happy, and how tormentingly she loved and hated him, and how terribly her heart was pounding. (Tolstoy 2001 [1878], 513–514)

There is obvious tension at the junction of the past and future. Anna simultaneously recollects past events and hopes for the future. Hope and hopelessness, anticipation of good and bad, sway steadily in her mind.

The human being lives in the space between the determined fate as a social being and is always in danger of losing individuality and dignity under the constant pressure of social convention and the expectations of respectability. In addition, resolution against this pressure in the form of unrestrained gratification of the spontaneous ego, and of freedom, seems to point to danger, discomfort and the ultimate and inevitable disaster of the individual. Thus, Anna truly is, as she says repeatedly, both guilty and yet not to blame. She is the tragic victim of human nature which calls both for the unhindered expression of the individual and the antithetical acknowledgment of the ultimate dependence of the individual upon the social.

Ernst Bloch (1995) argued that the human being is not yet fully determined. We all encompass a potential future self and this is the starting point for analysis, the route towards ourselves, the archeology of our self. The ontological meaning of unfinishedness or incompleteness assumes the form of longing for the totally other, the fulfillment of the human being. This longing is a central feature of the utopian impulse. Utopianism is a reaction to the felt incompleteness. Zygmunt Bauman (2003, 11) writes that the utopian 'urge to transcendence', rooted in our ontological incompleteness, is a 'constitutive feature of humanity'.

Paolo Freire (1998, 58) saw that 'it is in our incompleteness, of which we are aware, that education as a permanent process is grounded'. Our incompleteness grounds our educability and curiosity, our inquisitive yearning for something more; to be more than we are feeds the process of learning. If we think about teaching in this context, we see it as a call from an unfinished person towards other unfinished persons to reach wholeness together. An axiomatic starting point of education is to see it as action that grounds the very possibility for human beings to reach fullness, while the practical form of this call would be a space for human beings to understand their concrete incompleteness. It is only through this process, that we as human beings can attain the motivation to search for something that not yet is.

> If we reflect on the fact that our human condition is one
> of essential unfinishedness, that, as a consequence, we are
> incomplete in our being and in our knowing, then it becomes
> obvious that we are 'programmed' to learn, destined by our very
> incompleteness to seek completeness, to have a 'tomorrow' that
> adds to our 'today'. In other words, wherever there are men and
> women, there is always and inevitably something to be done, to
> be completed, to be taught, and to be learned. (Freire 1998, 80)

What we can become exists already in our present selves, but concealed behind the clamor of the given condition. The teacher works here as a utopian archeologist by trying to unearth utopian possibilities. Becoming more than we are requires the ability to maneuver through the noise of the present. The problem is that the language used in this process is the very same language that hinders the process. This forms the tragedy of utopia and the limitation of the possibility of utopia (Sargisson 2012, 39).

If the fulfillment of the human being is present, an opportunity emerges to begin loosening the chains of the given situation upon an individual. Each apparently airtight governing system is eventually very porous and has cracks if one actively searches for them (see Lakkala in this volume). As Bloch writes in the introduction to the *Principle of Hope*: 'The world is full of propensity towards something, tendency towards something, latency of something, and this intended something means fulfilment of the intending'. This dynamism is visible, as people act in the hope of so far unrealized human fulfillment. Our mind is set free in the new horizons opened by our daydreams. In these daydreams, utopia is transformed into the practice of outlining the not present in the given situation. 'But something's missing', as one of the main characters in Bertolt Brecht libretto to Kurt Weill's opera *Rise and Fall of the City of Mahagonny* (1930), insists while his friends continue to celebrate life.

The given situation pushes us towards the future unrelentingly, but our sight is fixed on the past. We stand upon the apparently full, but in the end empty present, because what is missing in the present is hidden underneath the material conditions and ideological justifications of the given conditions. Hegemonic ideologies change futures to eternal repetitions of the present. Variation might exist, but the key to harmony stays constant. This situation is further emphasized when disintegrative processes reach the anticipatory consciousness in education and our daily lives. Henry Giroux (2013) calls these cultural processes and functions of public schooling the 'disimagination machine':

> The 'disimagination machine' is both a set of cultural apparatuses extending from schools and mainstream media to the new sites of screen culture and a public pedagogy that functions primarily to undermine the ability of individuals to think critically, imagine the unimaginable, and engage in thoughtful and critical dialogue. Put simply, to become critically informed citizens of the world. (Giroux 2013, 263)

This disimagination machine collapses the future in to the present and makes it difficult or even impossible to think beyond the existing society, be it in terms of good society, the just distribution of wealth, etc. The horizon is blocked by the existing construction.

Max Horkheimer, a key member of the so-called Frankfurt School, wrote in his book *Eclipse of Reason* (1947) about how the collapse of the Enlightenment project of objective reason has opened the way for a society ruled by subjective reason. Horkheimer characterizes subjective reason as the reason of formal mathematics and scientific thinking the content of which is seen as politically and morally neutral. This is instrumental thinking in its purest, most rigorous form.

> The present crisis of reason consists fundamentally in the fact that at a certain point thinking either became incapable of conceiving such objectivity at all or began to negate it as a delusion. This process was gradually extended to include the objective content of every rational concept. In the end, no particular reality can seem reasonable per se; all the basic concepts, emptied of their content, have come to be only formal shells. As reason is subjectivized, it also becomes formalized. (Horkheimer 2013 [1947], 4)

Horkheimer argues that in the history of philosophical and scientific reasoning, those who have adopted objective reason have considered themselves determining the fundamental ends. In a sense, they have claimed that human reason has scope over given conditions and situations by stressing the teleological concerns of humanity. Even though Horkheimer was critical towards the philosophies of objective reason, he saw the responsible application of this kind of reasoning to allow us to find what is good, and what some particular goal ought to be. Subjective reasoning is more limited in scope. Rather than identifying ends, it merely 'coordinates' the appropriate means with some given ends in a given system. In other words, the question turns from 'What constitutes a good life?' to 'If I wish to accomplish x, what line of action must I take?'

Utopian anticipatory thought places itself between objective and subjective reason. It forms a link between the given and the hoped.

The anticipatory consciousness is connected with the remembrance of the past, with many having a sense of a repressed past. Walter Benjamin wrote in his famous *Theses on the Concept of History* (1940) about now-time which contains the dynamic relationship between past and future. The role of thinking is to explode the present as possibilities of something other than the given. Utopian pedagogy can be such an act.

### From Utopian Pedagogy Came Forward Movement

The term 'utopian' is often taken to refer to unrealistic ideas and improving the world in a way which will prove unsustainable, when tested against the real world. We argue that utopian thinking both draws upon and generates ideas capable of informing educational practices. Utopian thinking incorporates 'social dreaming' (see Introduction and Laakso in this volume), the complex of 'dreams and nightmares that concern the ways in which groups of people arrange their lives and which usually envision a radically different society than the one in which the dreamers live'.

Teachers have an enormous task in their hands. They must help prepare a new generation to live in a so far non-existing world. This means that teachers should enable and encourage students to take an active role in the construction of future society. Only this activity animates the world to be what it is hoped to be. This active role requires the simultaneous skills of imagination and active participation. In a sense, we all live in a mirrored room. Constant reflections from walls covered by mirrors makes the room intriguing yet strangely frightening at the same time. We can be aware of what is and how we have reached the present, but not what will happen in the future. The gap between continuity and change is constantly present in the life of a teacher.

Freire (1972) initially coined the notion of utopian pedagogy in *Cultural Action for Freedom*. Utopian pedagogy is a form of pedagogy, which is both highly critical and oppositional towards the given, and thus opens up a forward-looking horizon of new possibilities. Criticism is connected with a 'utopian vision about the human being and the world' (Freire 1972, 40). This practice is deeply dialogical in nature. In a pedagogical dialogue, the teacher and the student investigate together existing inhuman tendencies and proclaim the change in the name of humanity. In this practice, utopia is a reflection of where we are now. Without utopias, we would not know where we are standing and where we should be going.

Utopian pedagogy begins with the archeology of consciousness. The aim of archeology is to unearth the hidden and repressed (Freire 1981, 58). Fredric Jameson (2005) writes about archeology of futures and Ruth Levitas (2013) define archeology as the main form of the utopian method. The future manifests itself as ruins. Among these ruins of the past, the given moment and the hopes of the past and present human beings, an educational archeologist connects the clues to be able to find the blurred outlines of possible futures. Utopias have many faces. It is upon the observer to decide if utopia is the representation of the desirable dream or a daydreaming, unattainable dream or something in between.

To get back to Jameson, we can summarize his project in two very short statements. On the one hand, 'we have to name the system' (1995, 418); on the other, we have to *learn how to imagine utopia* (1988, 355). In the first case, there is the problem of mapping the totality of social relations within the disorienting conditions and political uncertainty of late-stage capitalism.

An aesthetic of cognitive mapping – a pedagogical political culture which seeks to endow the individual subject with some new heightened sense of its place in the global system – will necessarily have to respect this now enormously complex representation dialectic and invent radically new forms in order to do it justice. (Jameson 1995, 54)

For Jameson the map becomes pedagogical at the point when the individual situates the local in relation to the global. Such a map gives understanding to contradictions within capitalism without ideologically resolving these contradictions. This kind of cognitive map speaks not simply to the learning of new content, but rather, the process of learning to learn in a new way antithetical to standardization and to thinking, which parallelize reason with reality, and fact with the absolute.

In the second case, we face the question of envisioning a different world within a world that is hermetically closed to promote the given.

Analyzing the material and political preconditions for utopian imagination, Jameson (2004, 45–46) suggests that utopias are written during

> periods of great social ferment but seemingly rudderless, without any agency or direction; reality seems malleable, but not the system; and it is that very distance of the unchangeable system from the turbulent restlessness of the real world that seems to open up a moment of ideational and utopian-creative free play in the mind itself or in the political imagination.

### From Despair to Hope: Utopian Education in Practice

Jameson's two points – (1) *we have to name the system* (1995, 418) and (2) we have to *learn how to imagine utopia* (1988, 355) – resonate strongly with the objectives of teachers' work in schools. Two temporal aspects, present and future, are at the core of their work, but the orientations towards this can vary a lot in systems where teachers and schools have an autonomous position, like in Finland. Accordingly, orientations vary from the attitude where the world is 'taken for granted' to a critical approach towards work and society, including utopian thinking. In Finland, teachers live between tensions, where the curriculum strengthens them to imagine the future, and at the same time, practical political solutions force them to take into account demands coming from the labor market or politicians in power. This creates a visible conflict: How is it possible to seek a better world without hegemonies in a world based on hegemonies, seems to be a perennial wicked problem in education (Värri 2019). Nevertheless, the autonomous educational system and trust bestowed on teachers create a frame, where utopian thinking and implementation is possible. Next, we will introduce a concrete example of a school community practicing utopian education, beginning with its context.

Post-World War II Finland was an agrarian republic with deep class divides, which also reached into the field of education. The educational system was unequal, and secondary school was expensive and therefore inaccessible for most children from poor families. Since the 1950s, the system was vocally criticized especially by left-wing parties, and in the 1960s and 1970s, a radical educational reform was implemented as part of the welfare state construction process. In the process, the parallel school system (where some of the students applied to grammar school after 4th grade) was replaced by a nine-year comprehensive school. In addition, the social security system was reformed to support educational equality. Education was no longer dependent on the student's family background and wealth.

Huge structural reforms in education continued at the turn of the 1980s and into the 1990s, when the existing school inspection system was dismantled. In practice, schools became highly autonomous units in terms of both pedagogy and the operative culture, in the frame of a national core curriculum. Since the early 1990s, Finnish education has been based on trust among stakeholders in society, because the national core curriculum is loose and thus enables different interpretations to be made by teachers in the school community. The ethos of the curriculum and developmental work in education emphasize teachers as intellectual communities responsible for carrying out the objectives written in the content of curricula since the 1970s. However, the big question is: Are teachers and schools living and implementing this utopia in their everyday life?

We have met dozens of teacher communities in different contexts: in-service training, shared teaching experiments and research collaboration. Conversations with them have been similar year after year. Teachers are satisfied with their autonomy and pedagogical freedom, and very committed to their work. On the other hand, teachers feel that their work is too individual. They wish to have more co-operation, conversations and time with their colleagues in their everyday lives. The tension between individuality and communality is an integral part of teachers' communities and their day-to-day-lives.

The initial teacher education emphasizes critical and reflexive attitudes and skills, trained mostly via group-work (see e.g. Curricula 2014–2017). Why, then does this kind of professional development not continue in schools? Why are teacher communities in schools not constructed around critical, communal reflection, for example in the sense of collective, critical reflection defined by Suoranta and Moisio (2006). We argue that:

1    Teachers avoid risk-taking, to challenge the tradition of individuality in teachers' work (teacher as the individual survivor).

2   Conditions at the school level do not support communality as a basis for teachers' work.

3   Change is technically simple, but socially complex (Fullan 1982). The creation of a new culture requires a lot of common value-making through discussion and working together. This demands a lot of time as well as openness towards new thinking.

As stated earlier, Finnish schools are autonomous units. In addition, schools are like mirrors for themselves. They are similar all around the country, even if the differences have been growing during recent years because of various factors like educational leadership, different teaching and learning experiments, and collaboration with other stakeholders in society.

We reflect our three arguments mentioned above in the context of one school, which has undergone huge change during the past three years: from despair to hope, from individuality to communality and from strict structures towards flexible structures. In other words, the school represents utopian education in practice. This school under study is located in Sannainen, near the old medieval city of Porvoo, 50 kilometers east from Helsinki. It is a Swedish-speaking rural primary school, with eight teachers and six classrooms. Teachers initiated radical change in 2017. The reason behind the change was despair. Teachers were extremely exhausted. Many of them felt powerless in front of pupils, who were misbehaving. Teachers worked without interaction with colleagues, sick leaves were frequent and angst towards everyday life in school and the teaching profession grew day by day. Matti Rautiainen interviewed teachers in Sannainen School in the autumn of 2017, and again in 2018. The following analysis is based mostly on this data, and visits to the school including observations in classrooms.

In 2017, a U-turn was made among teachers. According to many teachers, they faced a wicked problem. They understood that they could not continue working the way they did, yet couldn't see or find any solution to the problem, because it couldn't be solved on the level of individual action. Problems concerned the whole school, not only one classroom. Finally, the school's management was re-organized: A new pedagogical leader was appointed. She was a teacher who had experience and interest to develop the school according to principles of the new curriculum. The aim of the work of the pedagogical leader was to supervise teachers towards a community-based learning organization. This was a new role, as schools do not usually have a specific pedagogical leader. In other words, a new agent in the school brought hope for the teachers.

What did the new pedagogue bring to the school then? The national core curriculum for basic education emphasizes active and inquiry-based learning processes, stronger student participation in schools and stronger interaction with other stakeholders in society. Since the early 1970s, when the first curriculum for the nine-year long comprehensive school was written, curricula have represented a utopia of an equal, democratic and humane society reflected in education and schools. Curriculum after curriculum, these themes have been restated and emphasized, but at the same time schools are far from realizing these big objectives. The pedagogical leader brought these big ideas and questions back to the community of teachers. Hope already existed in the school, but teachers had forgotten it.

The change that the pedagogical leader offered for the teachers was based on the idea of collaborative work around the curriculum and theme-based learning in classrooms. She asked: Could such a school be conceivable, where learning would follow themes rather than subjects? Could it be possible to build a strong collaborative culture between teachers? And finally: Could there be more space for the voice of the pupils in the school? The first reactions from the teachers were astonishment, signaling incomprehension and disbelief. However, because of the prevailing desperation, the new pedagogical leader was given the possibility to explore these questions. The teachers thought that the situation could not get any worse.

During 2017, the teachers gathered, every week, for hours, to read the curriculum and discuss how they understood not only the contents, but also the spirit of the curriculum. Week after week, teachers became more and more interested in new ideas that they developed together. Finally, the teachers created a new interpretation (utopia) on what their primary school should be like, and began to implement their new model in the following autumn. In our interviews, they all described a similar kind of developmental process, in which they first were astonished by the ideas presented to them by the pedagogical leader, but later became critical towards their own former professional identity and the way of teaching. Shared work and experiences were strongly emancipatory by nature. It changed the professional identity of most of the teachers in Sannainen.

As one teacher testifies:

> When I look at myself, and I think what I did before the spring of 2017, I look like a fool, a blind man, who was happy for the work he did, but did not understand at all the nature and deeper meanings of the profession. I could not believe anymore in teaching in the same way I worked earlier.

The process created 'a flow' amongst teachers, indeed they use the word 'flow' to describe the state of their own community. In the beginning of this chapter, we put forward three arguments on what prevents teacher communities from becoming collective and critical, with an ability to undertake utopian education. In the case of Sannainen:

1   Teachers avoided risk-taking, and they could not see any other way of schooling but the way in which they were socialized. They needed a critical voice from outside the community to reveal this. Teachers were open because they were in a desperate situation and a new vision represented a future hope instead of a threat.
2   Sannainen School had a pedagogical leader, who supervised and supported the process of change. In other words, the pedagogical leader created space and time for teachers to gather and discuss and made utopian education possible.
3   Teachers were open to the new ideas and were motivated to work hard towards their utopia as a teacher community. They constructed an intellectual community by themselves, trusting their capacity to work as a creative professional unit.

From these points of view, we want to ask three questions, closely related to utopian education as well as constructing collective and critical teacher communities.

Is the state of despair among teachers a requirement for utopian education?

Is utopian education possible within the frame and conditions of contemporary school structures?

How can teachers emancipate and empower themselves for utopian education?

Openness and curiosity towards a new direction in the future were also mentioned by teachers in the interviews as key concepts in professional development. Curiosity is also at the core of Paulo Freire's utopian education:

> The epitome of negation in the context of education is the stifling or inhibition of curiosity in the learner and, consequently, in the teacher too. In other words, the educator who is dominated by authoritarian or paternalistic attitudes that suffocate the curiosity of the learner finishes by suffocating his or her own curiosity. (Freire 1998, 79)

### Dynamic Connection of Hope and Despair

In a sense, hope and despair are two sides of the same coin. This connection between hope and despair is as natural as day and night, death and life, or happiness and sorrow. One cannot have one without the other. Erich Fromm analyzes this dynamic connection of hope and despair in the penultimate chapter of his book *You Shall Be as Gods* (Fromm 1966, 201–223).

In the chapter on 'Psalms', Fromm tells that after the destruction of the Temple, Psalms became the most popular prayer book among the Jews. The psalms ceased to be part of the Temple ritual practices and assumed a new function in this new historical situation. They became a human document, the expression of man's hopes and fears, joys and sorrows. They transcended the particular conditions of time and religious dogma and became the intimate friends and companions of Jews and Christians over many generations.

Fromm divides the psalms into one-mood psalms and dynamic psalms. It is the nature of the dynamic psalms that he wishes us to appreciate. They reflect a change of mood in the poet.

> The essential feature of the dynamic psalm consists in the fact that a change of mood is going on in the poet, a change that is reflected in the psalm. What happens is that the poet begins the psalm in a mood of sadness, depression, despair, or fear; usually, in fact, it is a blend of these various moods. At the end of the psalm his mood has changed; it is one of hope, faith, confidence.

Often it seems as if the poet who composed the end of the psalm was a different man from the one who composed the beginning. Indeed, they are different, yet they are the same person. What happens is that a change has occurred within the Psalmist during the composition of the psalm, he has been transformed; or better, he has transformed himself from a despairing and anxious man into one of hope and faith. (Fromm 1966, 207)

Fromm uses the word 'dynamic' to indicate that the change arises from the poet's struggle with his sense of despair. The struggle begins with an expression of despair, changes to a mood of hope, and then returns to deeper despair. Hope returns, only to plunge into deepest despair; and only at this point real hope emerges. Fromm uses this discussion of the dynamic psalm to convey his belief that only the person who experiences the full depth of his despair can liberate himself from despair and achieve hope (cf. Pekkola 2010, 139).

As argued by Sigmund Freud (2003 [1920]) in *Beyond the Pleasure Principle*, we all long for an initial state. In this initial state, someone else bears all the responsibility, in the most radical sense even thinks on our behalf. However, the road that leads to autonomy is the road that takes us further away from this primordial state of oneness. In a sense, it starts when we suddenly see that not everything that we value was presented to us. I was not there. This one sort of alienation, or better still objectification, where we start to gain a foothold outside the miracle circle of the initial state, when we develop our self-consciousness, begins when there are enough positive experiences of our own possibilities and strengths. In this development of autonomy, the other is the key that opens or locks the doors. Nevertheless, our longing to return to the womb of the initial state never leaves us, and thus we live in a constant state of tension between the initial state and something that waits outside.

We want freedom and autonomy, but at the same time long for shelter, intimacy and the atmosphere of trust. But the dread of losing ourselves into an involuntary and numb state of being drives us as an individual forcefully forwards. Capitalistic society teaches us to be careful not to lower our defenses, so that our vulnerability or defenselessness becomes known to others. Vulnerability, which is a sine qua non, is not a sufficient reason for trust. Trust is needed for autonomy to emerge.

Utopian thought informs educational and social practices: It enables processes whereby intentional communities determine material practices; it shapes visions for improved world orders; and it pervades cultural production that engages with utopian and dystopian ideas. We should rehabilitate Friedrich Schiller's (1965 [1795]), Herbert Marcuse's (1998 [1955]) and Fredric Jameson's (2005) emphasis on play, for it is through the apparent political paralysis of the present that the mind is, ironically, liberated to engage in utopian dreaming beyond given historical and political conditions. It is for the radical utopian pedagogy to give a certain discipline to this spontaneous, collective, timeless play, for without such direction the play will lose its radical dimension and fail to fulfill the potential of our utopian hope.

## References

Bauman, Z. (2003) 'Utopia with No Topos'. *History of the Human Sciences* 16 (1), 11–25.

Benjamin, W. (1969 [1940]) 'Theses on the Concept of History'. In W. Benjamin, *Illuminations*. New York: Knopf, 257–258.

Bloch, E. (1995) *The Principle of Hope*, Volume 1. Cambridge, MA: MIT Press.

*Curriculum Plans 2014–2017*. Department of Teacher Education, University of Jyväskylä.

Freire, P. (1972) *Cultural Action for Freedom*. Harmondsworth: Penguin.

Freire, P. (1981) 'Education for Awareness: A Talk with Paulo Freire'. In R. Mackie (ed.) *Literacy and Revolution: The Pedagogy of Paulo Freire*. New York: Continuum, 57–69.

Freire, P. (1998) *Pedagogy of Freedom*. Lanham, MD: Rowman and Littlefield.

Freud, S. (2003 [1920]) 'Beyond the Pleasure Principle'. In S. Freud, *Beyond the Pleasure Principle and Other Writings*. London: Penguin, 43–102.

Fromm, E. (1966) *You Shall Be As Gods: A Radical Interpretation of the Old Testament and Its Tradition*. New York: Holt, Rinehart and Winston.

Fullan, M. (1982) *The Meaning of Educational Change*. Toronto: Ontario Institute for Studies in Education.

Giroux, H. (2013) 'The Disimagination Machine and the Pathologies of Power'. *Symploke*, 21 (1–2), 257–269.

Horkheimer, M. (2013 [1947]) *Eclipse of Reason*. New York: Bloomsbury Academic.

Jameson, F. (1988) 'Cognitive Mapping'. In C. Nelson and L. Grossberg (eds.) *Marxism and the Interpretation of Culture*. Chicago, IL: University of Illinois Press, 347–360.

Jameson, F. (1995) *Postmodernism or, the Cultural Logic of Late Capitalism*. Durham, NC: Duke University Press.

Jameson, F. (2004) 'The Politics of Utopia'. *New Left Review* 25, 35–54.

Jameson, F. (2005) *Archaeologies of the Future: The Desire Called Utopia and Other Science Fictions*. London: Verso.

Kropotkin, P. (1974 [1912]) *Fields, Factories and Workshops Tomorrow.* London: Freedom Press.

Levitas, R. (2013) *Utopia as Method: The Imaginary Reconstitution of Society.* London: Palgrave Macmillan.

Marcuse, H. (1998 [1955]) *Eros and Civilization: A Philosophical Inquiry into Freud.* Boston, MA: Beacon Press.

National Core Curriculum for Basic Education (2016). Helsinki: Finnish National Agency for Education.

Pekkola, M. (2010) *Prophet of Radicalism: Erich Fromm and the Figurative Constitution of the Crisis of Modernity.* Jyväskylä: University of Jyväskylä.

Sargisson, L. (2012) *Fool's Gold? Utopianism in the Twenty-first Century.* Basingstoke: Palgrave Macmillan.

Schiller, F. (1965 [1795]) *On the Aesthetic Education of Man: A Series of Letters.* New York: Frederick Ungar.

Suoranta, J. and Moisio, O.-P. (2006) 'Critical Pedagogy as Collective Social Expertise in Higher Education'. *International Journal of Progressive Education* 2 (3), 47–64.

Tolstoy, L. (2001 [1878]) *Anna Karenina.* London: Penguin Books.

Värri, V.-M. (2019) *Kasvatus ekokriisin aikakaudella.* Tampere: Vastapaino.

PART III

**PERSPECTIVES ON UTOPIAS**

# 7 | THE SIGNIFICANCE OF HUMOR AND LAUGHTER FOR UTOPIAN THOUGHT

*Jarno Hietalahti*

Imagine a superior society.[1] Is it a peaceful place? Are people living in harmony? Are there no wars? These are typical features related to utopian worlds. However, people seldom ask, if there would be humor and laughter in such a perfect society. Despite the supposedly innocent character of humor, the question is important because the given answer reveals something essential about humanity and how it is conceptualized.

This chapter analyzes humor and laughter in relation to utopian thought. This is done by combining the leading theory amongst humor researchers, the incongruity theory, with the notion of utopia as a method of imagination. The relationship between utopia and humor is a rarely studied subject, as most often utopia is merely seen as a boring and humorless place. In opposition to this common standpoint, it will be argued that there are at least three different ways in which humor is related to utopia:

1 Laughing at utopia
2 Laughing with utopia
3 Laughing in utopia

In this chapter, the term 'humor' is treated as an umbrella concept, covering various genres of humor from farce to satire, from irony to buffoonery, and from slapstick to witticism. On a theoretical level, humor is based on a contradiction as stated by the incongruity theory: A classic example would be a man wearing women's clothing or a human being acting like an animal. This theory covers the whole field of human action and thinking; a person slipping on a banana peel is humorous, because normally people should be aware of their surroundings, and similarly a professor forgetting where they put their glasses while they are resting on their forehead is humorous, because a bright mind should be able to handle simple matters of their everyday life. Roughly, humor occurs when expectations do not meet the actual occurrences in the world. Furthermore, nothing is humorous in itself, but only in comparison to something else; a camel is not humorous on its own but only when it is compared to, for instance, another animals, or placed in a peculiar setting, like in a joke about drinking whiskey in a bar. To summarize, humor is a relational subject matter (see Raskin 2008).

Laughter, for its part, is understood as a reaction to humor, and not, for example, as a reaction to tickling or triggered by intoxication. On the simplest level, laughter expresses that the subject has perceived something ridiculous about the object/situation, which is treated in a humorous manner. It is amused, mirthful or joyous laughter which does not always require a physical expression – people can laugh silently in their hearts, so to speak. This position is based on Helmuth Plessner's (1970) idea that laughter has an expressive character; what laughter expresses is not fixed and universal (e.g. 'laughter is always a sign of happiness' is not so, as it may reveal embarrassment), but it nevertheless expresses something, and this something is connected to the social values and cultural categorizations which are shared in a society or culture. Human beings are always located in a specific historical period, which gives structure to every single individual, and *vice versa*, a society is always formed by individuals. On this deeper level, laughter expresses, as Plessner puts it, the human condition in the world.

As noted, utopias can trigger laughter. In the light of the incongruity theory, considering a utopia ridiculous means seeing something contradictory judging by the criteria of the prevailing social system in the utopia in relation to the prevailing social system. This ridiculous aspect is in some way inferior to what is perceived as the normal situation, and therefore undesirable. Laughter *at* utopia is conservative. Laughing *with* utopia, in contrast, accepts the deviation from the existing society as articulated by the utopia. This kind of humorous but critical way to relate to utopia is based on affirmative laughter; laughter confirms that there is something wrong with the prevailing system, and through utopian thought it aims to change the undesired social features. The third category, laughing *in* utopia, moves beyond the binary division of critical/conservative humor. This kind of humor challenges the prevailing cultural basis of current humor;

it offers something completely different. This category can be further divided into *humor within utopia* and *a utopian kind of humor*. I argue, that laughing in utopia requires that humanity (and humor within it) is necessarily in a state of fluidity. This type of humor is shaking and disturbing, hilarious and scary.

The three-fold distinction put forward here helps to understand the relationship between humor and utopian thought in a more profound manner than in previous studies, and it opens up new possibilities for both research on utopias and humor studies.

## Laughing at Utopia

It is easy to understand why utopias can be considered silly places, at least if they are seen as blueprints for better societies (see the Introduction of this book). They portray a picture of a place that is in many ways oppositional to the present society. Be it the biblical Garden of Eden, Plato's *Republic* (381 BC), Thomas More's *Utopia* (1516) or Francis Bacon's *New Atlantis* (1626), the outcome is often an unrealistic dream which in some sense or other is ridiculous. The eternal peace in the Garden of Eden is impossible because human beings are not always peaceful; the Republic with its strict rules considering, for instance, comedies and laughter, cannot be realized because human psyche does not work in the demanded way; Utopia will collapse because of its absurd politics; and New Atlantis will eventually sink when scientists try to act as leaders. These kinds of blueprint utopias will not work because they are ludicrous and impossible dreams.

Philosophy of humor helps to clarify why people laugh at utopias. A sense of humor is an essential human trait, and it guides how people deal with incongruities and surprises. According to the incongruity theory (see Raskin 2008), humor is based on paradoxes that need to be solved. When cultural conceptualizations are in contradiction, people have to be able to make some sense of the perceived oddity. For instance, if someone tells a joke: 'Two goldfish are in a tank. One looks at the other and says: I'll drive, you take the guns', one has to be able to consider the double meaning of the word 'tank'. If the keyword, tank, would have been, say, bowl, the joke would have been altogether different.[2] In this version, tank can refer both to an army vehicle and to a water container. Getting the joke (whether you consider it funny or not) is a sign of the flexibility of the human mind. Humans can play with conceptualizations and categorizations. If people stuck to the exact meaning of every word, their outlook on life would be much more restricted because there would be no room for paradoxes. Flexibility enables human beings to look beyond their current condition. In humor, things are different – and this can be a positive or a negative state of affairs for the individual perceiving the oddity.

Scholars of humor such as Aarne Kinnunen (1994) and Charles Gruner (1997) point out that the ridiculed target cannot be taken seriously. If the target happens to be a person, this target will no longer be highly esteemed.[3] Even if a joke, a caricature or a situational comedy focuses on, for example, a person's particular feature (e.g. oversized nose, old-fashioned style, stammering), ridicule functions as a stigma and questions the potential seriousness of the whole personality. Cruelly enough, if your hair is funny looking and others mock you because of it, it is hard to change the minds of the deriders with any kind of intellectual argument about your otherwise deep and virtuous characteristics. The easiest solution for the weak-minded is to have a proper haircut, which other people will accept. Because of the social nature of humor, laughter offers a potential challenge for every kind of deviation from the norms.

According to Henri Bergson (1914), laughing at different kinds of aberrations is social bullying. He argues that laughter punishes those who behave in ill-mannered or unsuitable ways. This kind of laughter both punishes and encourages you to change your behavior. You make the mistake of wearing a pink shirt at the factory, and your co-workers will most definitely make the corrective gesture – mocking laughter – so that you will never repeat the mistake. According to the superiority theory of humor, people laugh when they notice some eminency in themselves in comparison to others (see Morreall 1987). Thomas Hobbes (1962 [1651]) claims that this reaction stems from a 'sudden glory', and it is a sign of an individual's superiority to others. Following this theory, laughter expresses disbelief and scorn towards ridiculous sights.

Laughter can in a similar way be a means of expressing doubt towards utopian thought. If one argues for utopian possibilities like Ruth Levitas (2013) and considers it a realistic alternative for the current social form of living, she or he is dealing with the core of humor. A utopist expresses something different from what others are used to hearing. As the incongruity theory states, the offered vision is in contradiction with so-called normal ways of living. Utopia is a better place, that is not here (see Introduction of this book), and thanks to this distance to the surrounding world, humor is, at least potentially, present instantly – and the vision for the better world is always in danger of being ridiculed in the way described above.

The offered utopian model attacks the prevailing way of life, and this rarely happens without contest – there are always people who are satisfied with the existing state of affairs, and of course those who may be dissatisfied but are afraid of possible change. Capitalism is a great example of this. It offers a good enough habitat for a relatively large group; because of this, for example, socialist or communist alternatives are not attractive options to the prevailing order. If a theorist or a revolutionary comes up with a socialist dream state and offers it to the people in capitalist societies, the alternative possibility will most likely encounter scorn and laughter. Laughing at utopia rips apart the potential seriousness of the alternative and strengthens the prevailing order.

A historical example serves as an illustration. In the 1600s, Galileo Galilei challenged the geocentric model of the world by claiming that the Earth was not the center of the universe, but revolves around the Sun instead. He based his argument on the new scientific evidence gained with, for example, the use of a telescope. Galileo's position was strictly against the prevailing order, and the church could not accept it. For some 1500 years, the Bible had taught that the Earth stood still and the objects in the sky revolved around the Earth. There was strong evidence for this position in the holy book, and the heliocentric view was condemned, as were those who defended this position. (For a detailed view on the controversy, see Heilbron 2010.) There are some pieces of historical evidence which suggest that Galileo was seen as a fool because of his views in physics (Bethune 2007 [1830]). Fools, in general, are ludicrous and a constant target of laughter.

This kind of conservative laughter laughs at absurdity. This laughter underlines the strength and significance of the prevailing social circumstances for the people living in them. If one has always lived in a culture which repeats the idea that the Sun revolves around the Earth, the very basic worldview is built on this idea; it is hard to see how things could be otherwise. In Galileo's time, a member of the church could hardly understand how this position could be wrong. Galileo challenged the medieval understanding of the universe, which made his ideas seem absurd. Even if the church has nowadays accepted the Galilean position, in 1610 it was impossible to achieve a so-called objective position to solve the controversy. Galileo was not able to offer watertight proof for his claims, as he had only uncertain evidence. Because Galileo's suggestion appeared absurd to many, he was ridiculed for his claims. People laughed at his alternative reality and through laughter expressed their disbelief.[4]

Conservative laughter can also be used as a political tool. When a republican president of the United States of America ridicules and laughs at, for example, alternative politics offered by democratic politicians, he uses humor and laughter as means for his own politics; ridiculing the opposite aims at enforcing current political agendas (for a detailed view of political humor, see Hietalahti 2019). Reportedly, in Germany the right-wing party Alternative for Germany has developed this technique even further; during parliamentarian debates the members of the party attempt to drown out other members' speeches with coordinated laughter. This is accompanied with insults at opponents and jeers to party fellows. If, for instance, a left-wing or a Green parliamentarian tries to argue for humane immigration policies, the right-wingers attack with coordinated laughter (Witte and Beck 2018).[5]

Naturally, it is uncertain what the actual consequences of this kind of laughter are (see Kuipers 2008). Analogically with racist humor, laughter may strengthen the questionable attitudes behind the offensive joke, but it is possible that racist humor functions inversely and questions the shared racist attitudes (see Weaver 2011). Simply put, telling a joke does not automatically lead to the triumph of the joke-teller and their political agenda. Whatever the actual consequences are, laughing at utopia nevertheless reveals something about the attitudes of the laughers; they see something ridiculous in the alternative world. This type of laughter is conservative because it, at least in principle, strengthens the *status quo*.

## Laughing with Utopia

Even if laughter is often conservative, there is variety in guffaws. Various philosophers of humor (e.g. Hutcheson 2009 [1750]; Kant 1987 [1790]; Freud 1968 [1927]; Morreall 2009; Hietalahti 2016) have noted that even though absurdity makes people laugh, it is not necessarily triggered by the feeling of their own priggishness and superiority. Instead, a twist in conceptualizations may be enough in itself to trigger amusement. A change of perspectives can be also affirming, and it can agree with the expressed humorous alternative. In regard to utopian thought, this means that people can laugh and agree with the new perspective or suggestion for societal evolution. Utopia can be *amusing* in the positive meaning of the word.

For instance, Thomas More describes a place entirely different from England in his time, creating a tension between the dream-like island and the existing society. As described in *Utopia* (1516), England is a dreadful place full of inequality, suffering and hope-lessness. Utopia, in comparison, is an island where people happily flourish. They have the possibility to live their lives in peace and harmony without envy or fear of violence and criminality.

In relation to humor, More's description turns the tables: Utopia is the good place and England a country that can and should be improved socially, culturally and politically. Contemporary England, one may reason, starts to look like a rather silly place, when an alternative is presented. Presumably the commonly shared nonsensical situation can trigger laughter, and this laughter does not need to be condemnatory, but it can stem from possibilities to change the real world into a better place. Laughter with utopia is positive and affirmative laughter that supports the offered alternative. In this light, the satirical structure of More's Utopia is apparent (for a more detailed take on More's literary techniques, see Elliott 1963).

In the field of literature, satire offers a typical form of critical utopian thought. It ridicules culturally shared follies and vices by shaming corporations and government, indeed, quite often the very society itself. This genre has been present for millennia, as the authority of divine emperors and famous theologians has been satiri-cally mocked at least since Lucius Anneus Seneca's times (Kivistö 2016). By comparing Utopia to historical satirical masterpieces, Robert C. Elliott (1963, 321) convincingly shows that Utopia has both a critical attitude and a normative model. More criticizes his contemporaries and calls readers to laugh with his utopian model. Humanistic theorist Erich Fromm highlights this idea when he describes the shared insanity of the Western world. In an ironi-cal manner, Fromm discusses how for his contemporaries, heaven would look like a huge shopping center full of all kinds of shops and new gadgets. The individual would have a tremendous amount of money – and of course a little bit more than his neighbors – and he would just keep buying with his mouth widely open in wonderment (Fromm 2008, 131). From a humanistic standpoint, this is foolish. Fromm goes on and lists a bunch of modern paradoxes between humanistic values and perceived reality, and reveals how everyday practices are eventually ludicrous. This can trigger bitter but nevertheless hopeful laughter:

> Does it make sense to spend millions of dollars on storing agricultural surpluses while millions of people in the world are starving? Does it make sense to spend half of the national budget on weapons which, if and when they are used, will destroy our civilization? Does it make sense to teach children the Christian values of humility and unselfishness and, at the same time, to prepare them for a life in which the exact opposites of these virtues are necessary in order to be successful? ... Does it make sense that we live in the midst of plenty, yet have little joy? Does it make sense that we are all literate, have radio and television, yet are chronically bored? (Fromm 2006, 92–93)

Fromm's own ideal society is a sane society in which people can live in harmony with others and with themselves. This is possible when a society meets the existential needs of an individual; a society exists for its inhabitants, not the other way round. Fromm offers a detailed list of alternative ways to arrange a society beginning with new democratic practices and ways of distributing wealth and goods. According to him, this is a distant but real possibility (Fromm 2008). A reader agreeing with Fromm's vision may very well laugh delightedly with Fromm's utopian vision. This laughter expresses that the currently shared Western insanity portrayed by Fromm is ludicrous.

Fromm admits that there is no guarantee that his paradise-like society would be realized by following his suggestions, but for him, it is important to try to imagine a real alternative for current practices. Fromm's position is largely compatible with Ruth Levitas' (2013) understanding of utopia as a method of imagination. It offers a possibility to think beyond the prevailing social systems. Humor and laughter are instruments in this process. They both foster imagination and strengthen the offered possibilities. The more ridiculous the present-day social organizations and international institutions appear to be, the more there is room and possibilities for utopias. Laughter enhances the possibility of change.

Even if conservative laughter is aimed at absurdity, absurdity in itself is not necessarily something negative. Even if utopia may sound in some sense absurd, it does not mean that the offered alternative is worthless. This idea is present already in Erasmus of Rotterdam's *The Praise of the Folly* (1511). Erasmus argued that laughter can break down the strongest structures of reason, by which he refers to widely accepted social customs. According to Erasmus, a rigid (in his context, biblical) reason does not meet the standards of flourishing human life. Instead, humanity is built on both reason and silliness, both of which are necessary. Absurdity, then, is inherent in human life and it should not be pushed aside. Erasmus paints a vision in which foolishness thrives and funny-sounding ideas are not immediately rejected because they appear to be against the normal order of the world. Instead, foolishness equals openness to various possibilities. On this metalevel, the medieval and scholastic way of describing humanity and humans' place in the world – despite the somewhat stale style of scholastics – starts to look ridiculous. On the other hand, fools and folly make people laugh, but in the way that opens eyes for new possibilities. This laughter is laughing with utopia. It is critical laughter which aims at changing the prevailing social reality – be this aim conscious or unconscious. This laughter expresses that there is a strong yearning for something else than what is offered in current societal conditions.

## Laughing in Utopia

As seen above, laughter can be either conservative or critical in relation to the prevailing social setting: If laughter is connected to the wish to change the prevailing circumstances, it is critical. If it aims at preserving how things are, laughter is conservative (see also Kuipers 2008). Both forms of laughter are in a sense external to the offered utopia. They express an attitude against or for a different society. However, the binary distinction between conservative and critical laughter does not cover all the possible ways in which laughter and utopian thought are intertwined. The third possibility is laughing *in* utopia.

With this type of laughter, it is possible to overcome the rigid dichotomies people assume toward humor and laughter. Instead, it refers to the laughter that happens in an entirely different social situation. Because humor is by definition interpersonal and socially formed (dependent on cultural categorizations), it expresses shared worldviews of and in a society. Laughing in utopia, then, is a reaction to humor that occurs in a new world. This category is two-fold: There is *humor within utopia*, and a *utopian kind of humor*. First, I will handle humor within utopia.

As mentioned in the introduction to this chapter, it is a gross exaggeration to see utopias as necessarily boring, even if the claims are understandable. Thomas More describes his ideal island as a place where there are no opportunities for wickedness (More 2012 [1516]); this may sound rather dull to those who think that humor should always break boundaries and challenge morality. In this light, it is not surprising that Gregory Clayes concludes that a utopian society offers security, but 'is not really a fun place' (Clayes 2016, 16). Similarly, Arthur Schopenhauer has commented on another form of utopia, the Christian paradise, in a more drastic manner: 'after man had transferred all pain and torments to hell, there then remained nothing over for heaven but ennui' (Schopenhauer 1910, 402). Charles Gruner further claims that '(h)umor could hardly exist in this aggressionless, peaceful utopia' (Gruner 1997, 35).

In the hands of these thinkers, the idea that utopia is a boring place becomes a deep criticism; utopia offers security, but the cost of this is an aching boredom. As ennui is seen as worse than social injustice, utopias turn out to be ultimately dystopic (e.g. Moravia 1965).

This criticism is based on, first, the idea that humor and laughter are in themselves valuable (and the lack of them is unwelcome), and second, on a misunderstanding about the nature of utopia. The first ideal is arguably wrong, because fun and amusement do not justify anything in themselves – funniness and its social value has to be evaluated in relation to morality (see Hietalahti 2016), so I will focus on the second aspect in this last part of this chapter. Let us have a look at Thomas More's Utopia to clarify the misunderstanding.

Even if More's utopian island seems like a boring place to the modern reader, this does not mean that Utopia is a dull society. On the contrary, More describes various forms of entertainment and humorous occasions from which the Utopians gain pleasure.[6]

They have fun, even if the described forms of humorous entertainment would not be enjoyable for modern Western people. However, modern readers do not have a monopoly over fun; modern human beings cannot dictate the universal requirements for amusement and entertainment. Instead, they should be understood in the light of the different social conditions of the different society, that is, Utopia. If one believes More, Utopians are not bored, but clearly enjoy how they pass time.[7]

Besides descriptions of fun, there is a utopian element of humor in More's book; this can be found in the treatment of fools. More writes how Utopians 'take great pleasure in fools ... they do not think it amiss for people to divert themselves with their folly' (More 2012, 146). This idea is related to the basic mood of Utopians, as none of them should be too sullen or severe to enjoy fools' 'ridiculous behavior and foolish sayings' (More 2012, 146). It may sound striking that people laugh at, for example, mentally handicapped individuals, but this was actually a humane and progressive idea in More's time. He clarifies the moral aspect of this kind of laughter:

> If any man should reproach another for his being misshaped or imperfect in any part of his body, it would not at all be thought a reflection on the person so treated, but it would be accounted scandalous in him that had upbraided another with what he could not help. (More 2012, 146)

Clearly, there is room for humor in More's Utopia, and it has an obvious connection to morality. This is an imaginative idea if one thinks that humor is an opposite to a virtuous life. By combining amusement and morality, More challenges the general idea about humor and laughter in his own time – that is, how people laugh as described in the light of the superiority theory above. When imagining morally eminent humor, More is a humanistic thinker: He demands openness in relation to humor, and prefers laughter which does not mock or belittle. Laughter triggered by fools' sayings is not laughter at them; it is laughter that expresses openness of thinking and enjoyment of alternatives.[8]

More's approach to humor and laughter is intriguing, but remains at a somewhat simple level. In his treatment, both laughter and humor signal ethical values which can be understood from a modern perspective, too. However, they can be pushed even further; it is possible that humor and laughter express something which is incomprehensible to modern readers. Arguably, utopian humor should be inconceivable. If utopia offers an entirely different social reality, humor and laughter should express humanity within this context. Then, humor becomes imaginative and cannot be easily grasped with commonly shared conceptualizations and within what is perceived as the so-called normal social setting. This leads to the second part of the category, the utopian kind of humor.

The idea of possibly incomprehensible humor can be approached and clarified with help from Ludwig Wittgenstein. According to Wittgenstein, a word or a sentence can mean something if it is expressed within a 'language-game'. Language used in an ill-mannered way is incomprehensible, as the rules of the game are not followed (Wittgenstein 1986 [1953]). A sentence like 'Is more sophisticated than real or a clockwork?' does not make sense even if the words are familiar to the reader. However, it is logically possible that in some different form of human life the sentence would make sense – even if it is hard to imagine what such a context would be like. It would be an entirely different setting, and understanding people from that reality would be tremendously difficult for humans from another kind of social reality. Wittgenstein argues that people could not understand lions even if they could speak (Wittgenstein 1986, §327). Their entire life situation would be so different that human beings' cognitive capabilities just would not match theirs. Here lies the most extreme challenge for utopian thought.[9]

This brings us back to humor. As argued above, humor is triggered when something unexpected happens. The deviation may very well be an unusual linguistic expression. It is possible to create a sketch in which the above-mentioned sentence would be in some way sensible. Or in some peculiar comedy, people could use this kind of silly language, and the audience could understand the funniness of the non-grammatic style of word usage. Reportedly, Wittgenstein suggested that it is possible and even desirable to write a philosophical work in the form of a joke (Malcolm 2001). It is not entirely clear what kind of humor would be philosophically valuable, but supposedly it would be in some way or another utopian humor. This Wittgensteinian kind of joke-telling can be called a utopian kind of humor. However, in utopian thought it would be necessary to move beyond regular ways of joke-telling, which could mean getting rid of particular kinds of setups and punchlines as well as traditional themes like differences between sexes or stupidity of politicians. Utopian humor challenges both the form and the content.

Helmuth Plessner offers a fruitful theory of laughter which is useful to deepen this idea. According to him, laughter has an expressive character, even though the social significance of laughter is not clear. For Plessner, genuine laughter is stripped of external meanings, and it expresses the human position in the world. Laughter is not laughter at or with, but *in* a human body; yet the body is always located in a specific historical context (Plessner 1970). In utopia humor would not have the same burdens which it has in contemporary society (e.g. demand for funniness, recognizable form, etc.).

Because this utopian type of humor is so different from what people are used to in their shared comedies and jokes, it is hard to describe it in detail. The main principles of utopian humor are that it does not merely challenge the limits of behavior and morality, but the limits of humor themselves. Therefore, utopian humor most likely would not be offered in traditional forms (e.g. jokes, stand-up shows, sit-coms, etc.), but it would challenge existing formats. Philosopher Theodor W. Adorno sees this type of humorous and revolutionary power in the works of Samuel Beckett. According to Adorno, Beckett's books and plays are not merely absurd without any significance, because then they would be just trivial. Instead, Beckett's works are meaningfully absurd because they challenge the way people use reason and rationality in the Western world – they put the meaning on trial, as Adorno claims (Adorno 2002, 153). Simon Critchley has made a similar kind of analysis of Beckett, where he claims that Beckett's plays laugh at laughter – they question the whole nature of humor and fun (Critchley 2002).

If humor is about paradoxes and surprises, then it cannot offer sameness. Adorno is highly critical towards a culture industry which tries to enclose humor in a safe form (Horkheimer and Adorno 2002, 114). In this process the cultural products become dull and repetitive, and humor is stripped away from its very core: a peculiar kind of merry non-sense which challenges the way people usually see the world. Following Adorno, humor and laughter should not be anything fixed but expressions of human freedom and imagination. This utopian element is easily left aside if one considers humor as merely a tool for fun.

This type of utopian humor and laughter are close to what Friedrich Nietzsche calls 'the golden laughter'. It is laughter that is not bound to everyday morality, but resonates with the new humanity which has moved beyond conventional ideas about good and evil (e.g. Nietzsche 2016 [1883]; 2013 [1886]). In this new situation, laughter still has an expressive character, but it is basically impossible to evaluate this laughter, for instance, in the framework of traditional morality. It is laughter by a new type of human being who is not bound to the old ways of living. As it happens, Nietzsche would like to rank philosophers according to their way of laughing – the best ones would be those who are capable of golden laughter (Nietzsche 2013 [1886]).

From a contemporary perspective, humor in utopia is essentially ambiguous because it is born from different cultural categorizations and exists in a different social reality. Therefore, it can be unsettling, disturbing, funny and scary, and because of its potentially incomprehensible nature, it can appear as nonsensical to an audience from a different kind of society. But as Adorno reminds, there is a possibility for a utopian kind of laughter already in the present circumstances, even if what he calls culture industry tries to suppress humor and laughter into a product of simpleminded fun. There is an ongoing dialectic process that reveals something essential about the prevailing conditions of humanity: the prevailing 'false laughter' (Horkheimer and Adorno 2002, 112; see also Hietalahti 2017) signals the negative aspect of current unimaginative humor, yet glints of a utopian kind of humor remind us that things could be otherwise. The fantastic form and content of humor offer a challenge for both utopian thinkers and scholars of humor, as well as to contemporary comedians. Understanding humor in utopia requires imagination, which allows one to transcend the social expectations of the current society.

Understanding utopian humor requires openness to alternatives. Of course, for a comedian, this task is tremendously hard if one aims to create a new kind of humor: Current audiences are unlikely to regard it funny. But even if the masses do not laugh, there might be philosophical depth in this new kind of humor. Utopian humor challenges the ways in which humor and laughter can be understood in general.

## Endgame

So, what is the answer to the question given at the beginning of this text? Is there room for humor in the imagined perfect place? If the answer is negative, and the perfect world would lack humor and laughter altogether, the concept of the human being is altered considerably. But, if one agrees that humor is an essential human feature, laughter probably occurs in the perfect place. However, the trick is that the current forms of humor may not be very utopian. If one tries to force utopian humor to fit his or her own sense of humor or taste, something essential may be left aside. Preferably, one should push for radically different humor – even if it might be too strange to the current ways people assess funniness as such.

I have argued that utopian thought and humor are related in at least three different ways. First, people can laugh at utopias. This type of laughter is conservative, because it rejects the alternative articulated by utopia. Laughing in this sense is related to preserving the prevailing social conditions. Second, people can laugh with utopia. This laughter is critical, because it expresses a wish to change the *status quo*. Laughing with utopia is affirmative and it states that there is something wrong with the current society – and this something should be changed. Third, there is laughing in utopia, which is a two-fold category. Obviously, there is humor within utopia which refutes the idea that utopias are boring places. Furthermore, utopian kinds of humor are philosophically not only the most interesting, but also the most demanding.

Utopian laughter is not merely about preserving or criticizing: It expresses a new kind of humanity and a new kind of society. Because of this, utopian humor and laughter can be incomprehensible to contemporary people. Furthermore, utopias are not merely places, but methods of imagination. In the process of imagination, humor and laughter should not be neglected. They are pivotal human traits which in their own peculiar way express essential aspects of the prevailing and of an entirely different society. However, this expressive character is not fixed and rigid, but dynamic and fluid. Utopian thinking therefore demands openness to alternative forms and contents of humor.

## Notes

1 It is crucial to note that not all utopias are distant islands or 'places'. However, envisioning a different and better kind of place in comparison to the prevailing social order is a work of imagination. Therefore, utopia as a place and as a method of imagination come close to each other.

2 Obviously, when you have heard the joke dozens of times, you can try to add some absurdity with an alternative choice of words.

3 Evidently, the politics of humor are much more complicated than Kinnunen and Gruner claim; ridiculing can, for example, strengthen the power of a politician (see Kessel and Merziger 2012).

4 Obviously, the geocentric worldview was ridiculous to Galileo, which is revealed in his letters to Johannes Kepler. Galileo wished that they could laugh together at the remarkable stupidity of both the common herd and the philosophers who declined to look through his telescope: 'Oh, my dear Kepler ... what shouts of laughter we should have at this glorious folly!' (Bethune 2007, 29).

5 Apparently, the Green and left-wing parties have adopted a similar tactic by using taunts and guffaws to question the right-wing politics (Witte and Beck 2018).

6 I recognize that 'entertainment' and 'humor' are not synonyms; there are various forms of non-humorous entertainment as well as humor that is not a form of entertainment. However, on a general level, there is plenty of entertainment that is based on humor, and often humor is entertaining.

7 Analogously, it would be implausible to claim that, for instance, Mauritius is a boring place just because the ways of living and forms of humor on the island are (presumably) different from in the writer's home country, Finland.

8 It is worth mentioning that Thomas More had his own domestic fool, Henry Patenson, who was included in More's family portrait by Hans Holbein – a rare honor in that era.

9 Clearly, not all utopias are this remote from everyday practices; there are 'real' utopias which draw from current circumstances and try to offer something familiar but different (see Wright 2010).

## References

Adorno, T. W. (2002 [1970]) *The Aesthetic Theory*. Trans. R. Hullot-Kentor. London: Continuum.
Bacon, F. (2008 [1626]) *The New Atlantis*. Retrieved from Project Gutenberg on June 30, 2019 www.gutenberg.org/ebooks/2434.
Bergson, H. (1914 [1911]) *Laughter: An Essay on the Meaning of the Comic*. Trans. C. Brereton and F. Rothwell. New York: Macmillan. Retrieved from https://archive.org/details/laughteranessayoobergggoog/page/n8.
Bethune, J. E. D. (2007 [1830]) *The Life of Galileo Galilei with Illustration of the Advancement of Experimental Philosophy*. Retrieved from the Internet Archive on June 30, 2019 https://archive.org/details/lifeofgalileogaloobethrich.

Clayes, G. (2016) 'Introduction: Utopia at 500'. In K. Olkusz, M. Klosinski and K. M. Maj (eds.) *More after More: Essays Commemorating the Five-Hundredth Anniversary of Thomas More's Utopia*. Kraków: Facta Ficta Research Centre, 14–24.

Critchley, S. (2002) *On Humour*. London and New York: Routledge.

Elliott, R. C. (1963) 'The Shape of Utopia'. *ELH* 30 (4), 317–334.

Erasmus (1941 [1511]) *The Praise of Folly*. Trans. H. H. Hudson. Princeton, NJ: Princeton University Press.

Freud, S. (1968 [1927]) 'Humour'. In *The Complete Psychological Works of Sigmund Freud*, Volume 21: 1927–1931. Trans. J. Strachey. London: The Hogarth Press, 159–166.

Fromm, E. (2008 [1955]) *The Sane Society*. London and New York: Routledge.

Fromm, E. (2006 [1962]) *Beyond the Chains of Illusions: My Encounter with Marx and Freud*. New York and London: Continuum.

Gruner, C. (1997) *The Game of Humor: A Comprehensive Theory of Why We Laugh*. New Brunswick, NJ and London: Transaction Publishers.

Heilbron, J. L. (2010) *Galileo*. Oxford: Oxford University Press.

Hietalahti, J. (2016) *The Dynamic Concept of Humor: Erich Fromm and the Possibility of Humane Humor*. Doctoral dissertation. Jyväskylä: University of Jyväskylä.

Hietalahti, J. (2017) 'Socially Critical Humor: Discussing Humor with Erich Fromm and Theodor W. Adorno'. *Idéias* 8 (1), 87–108. doi:10.20396/ideias.v8i1.8649776.

Hietalahti, J. (2019) 'Carl Jung and the Role of Shadow and Trickster in Political Humor: Social Philosophical Analysis'. In C. P. Martins (ed.) *Comedy for Dinner and Other Dishes*. Coimbra:

Instituto de Estudos Filosóficos (IEF), 20–41.

Hobbes, T. (1962 [1651]) *Leviathan: On the Matter, Forme and Power of a Commonwealth Ecclesiasticall and Civil*. New York: Collier Books.

The Holy Bible (2016) Retrieved from Biblia.com on June 30, 2019 https://biblia.com/books/esv/Ex11.1.

Horkheimer, M. and Adorno, T. (2002 [1947]) *Dialectic of Enlightenment: Philosophical Fragments*. Trans. E. Jephcott. Stanford, CA: Stanford University Press.

Hutcheson, F. (2009 [1750]) *Reflections upon Laughter: And Remarks upon the Fable of the Bees*. LaVergne: Kessinger Publishing's Legacy Reprints.

Kant, I. (1987 [1790]) *Critique of Judgement*. Trans. W. S. Pluhar. Indianapolis, IN and Cambridge: Hackett Publishing.

Kessel, M. and Merziger, P. (eds.) (2012) *The Politics of Humour: Laughter, Inclusion and Exclusion in the Twentieth Century*. Toronto: University of Toronto Press.

Kinnunen, A. (1994) *Huumorin ja koomisen keskeneräinen kysymys* ('The unfinished problem of humor and comic'). Helsinki: WSOY.

Kivistö, S. (2016) 'Satirical Apotheosis in Seneca and Beyond'. *Helsinki Collegium for Advanced Studies* 20, 210–230.

Kuipers, G. (2008) 'The Sociology of Humor'. In V. Raskin (ed.) *The Primer of Humor Research*. Berlin and New York: Mouton de Gruyter, 361–398.

Levitas, R. (2013) *Utopia as Method: The Imaginary Reconstitution of Society*. London: Palgrave Macmillan.

Malcolm, N. (2001) *Ludwig Wittgenstein: A Memoir*. Oxford: Clarendon Press.

Moravia, A. (1965) *Empty Canvas*. Trans. A. Davidson. London: Penguin Books.

More, T. (2012 [1516]) *Utopia*. Retrieved from Open Utopia on June 30, 2019 http://theopenutopia.org/full-text/introduction-open-utopia/.

Morreall, J. (1987) *The Philosophy of Laughter and Humor*. Albany, NY: SUNY Press.

Morreall, J. (2009) *Comic Relief: A Comprehensive Philosophy of Humor*. Chichester: Wiley-Blackwell.

Nietzsche, F. (2016 [1883]) *Thus Spake Zarathustra: A Book for All and None*. Trans. T. Common. Retrieved from Project Gutenberg on June 30, 2019 http://www.gutenberg.org/files/1998/1998-h/1998-h.htm.

Nietzsche, F. (2013 [1886]) *Beyond Good and Evil*. Trans. H. Zimmern. Retrieved from Project Gutenberg on June 30, 2019 www.gutenberg.org/files/4363/4363-h/4363-h.htm.

Plato (2003 [381 BC]) *The Republic*. Trans. T. Griffith. Cambridge: Cambridge University Press.

Plessner, H. (1970) *Laughing and Crying: A Study of the Limits of Human Behavior*. Trans. J. S. Churchill and M. Greene. Evanston, IL: Northwestern University Press.

Raskin, V. (ed.) (2008) *The Primer of Humor Research*. Berlin: Mouton de Gruyter.

Schopenhauer, A. (1910) *The World as Will and Idea*, Volume 1. Trans. R. B. Haldane and J. Kemp. London: Kegan Paul, Trench, Trübner & Co. Retrieved from the Internet Archive on June 30, 2019 https://archive.org/details/theworldaswillan01schouoft.

Weaver, S. (2011) *The Rhetoric of Racist Humour: US, UK and Global Race Joking*. Farnham: Ashgate.

Witte, G. and Beck, L. (2018) 'The Far Right Is Back in the Reichstag, Bringing with It Taunts and Insults'. *The Washington Post*, June 8.

Wittgenstein, L. (1986 [1953]) *Philosophical Investigations*. Trans. G. E. M. Anscombe. Oxford: Basil Blackwell.

Wright, E. O. (2010) *Envisioning Real Utopias*. London: Verso.

# 8 | ARCHITECTURAL UTOPIAS AS METHODS FOR EXPERIMENTING WITH THE (IM)POSSIBLE

*Aleksi Lohtaja*

Architecture is often considered to be a form of art, in which ideological and utopian dimensions are intertwined. Compared to other forms of art, such as poetry, literature, music or painting with a utopian function to envision imaginary worlds which critically counter the present, architecture never entirely flees from current society and its material basis. Instead, its utopian aspirations are always materialized through a social and political framework and its restrictions. In this sense, architectural utopianism seems to manifest Karl Mannheim's (1936) seminal argument regarding the coexistence of utopias and ideologies: political ideologies are 'realized' versions of utopian aspirations. However, the extent to which this defines the emancipatory potential of architecture (or utopias in general), is debatable.

The relationship between utopia, architecture and ideology is central to cultural theorist Fredric Jameson in his essay 'Architecture and the Critique of Ideology' (Jameson 2000; for a broader contextualization of this text, see Lahiji 2012). In the essay Jameson distinguishes (1) a very critical reading on the relationship between architecture and ideology in the works of Italian architectural theorist Manfredo Tafuri (1935–1994) and (2) a more affirmative yet also critical approach to the same matter in the work of French philosopher Henri Lefebvre (1901–1991). For both Lefebvre and Tafuri, it is precisely the close connection between utopianism and ideologies that encloses architecture's role in society. However, even though these two disciplinary approaches to architecture share the idea that architecture reflects the broader conditions and ideologies of society, their conceptualizations of political uses and potentialities for utopias differ fundamentally.

Can these different positions be taken to clarify the contemporary use of the concept of utopia in political and social theory? This article attempts to understand, in which way architectural utopias can be taken as methods for socio-political change (as discussed in the Introduction, see also Levitas 2013; Moylan and Baccolini 2007). I propose that (architectural) utopias are not inherently political in an emancipatory sense, nor do they, however, function only as plain ideologies subordinated to reinforced political goals. Instead, I consider utopias as a way to mediate politics and configure another pathway for the future, which implies always a utopian horizon. Rather than thinking of this in an ahistorical manner covering architectural utopias in general, my special focus is a critical reading towards modernist architectural utopianism offered by Tafuri and Lefebvre. Utopian thought has of course always implied certain architectural aspects, but the modification of space as an inherently utopian act is situated in the legacy of early twentieth-century avant-garde and modernism supplemented by the post-World War I impulse to change society through architectural design. An impulse wherein utopia is not understood only as fixed space, as was the case in the architectural aspects of 'classical utopias', but also as the mode of an activity (Margolin 1997; Heynen 1999; Henket, Heynen and Allan 2002; Hvattum and Hermansen 2004; Miles 2004; Coleman 2012; Lahiji 2019).

In the first section, I will discuss the nature of modernist architectural utopias and their manifested tendency to think that 'society can be changed through design'. I will briefly discuss how this definition has been criticized especially from the point of view of such prominent utopian theorists like Ernst Bloch and Theodor W. Adorno, who share the idea that the political efficacy of architectural utopias, understood this way, is closer to ideological and even totalitarian political programs than societal liberation.

In the second section, I will outline a more politically nuanced critique towards the definition of modernist architectural utopianism from the point of view of its conformist engagement towards the prevailing society. Theoretically this can be traced especially to Manfredo Tafuri's reading that takes the afterlife of modernist utopianism as his point of reference. Tafuri's critical reading finds an empirical touchpoint in early twentieth-century architectural utopianism, such as Russian Constructivism, The Bauhaus, New Frankfurt, the housing policies of socialist Red Vienna, and other modernist architectural movements, that are often considered socialist or left-leaning attempts to design society beyond classical nineteenth-century capitalism. However, Tafuri saw their reformist attitude towards society failing to understand the simultaneous transformation of capitalism and for example how the ideas of socialist de-commodified housing were best realized throughout the reorganization of capitalism towards a Taylorist-Fordist mode of production. Here the utopian aspirations function as the reproduction of society – an aspect that is commonly referred to as ideological, especially in Marxist scholarship.

Tafuri's Marxist reading of modernist utopianism seems to reflect the general tendency to consider 'utopian' and 'Marxist' approaches as opposite poles. Here, however, especially Lefebvre's discussion of architectural utopias, examined in the third section, from another Marxist perspective offers another approach to the problematic relationship between utopia and ideology. While Lefebvre's theory has significant similarities with Tafuri's as regards the ideological dimension of modernist utopianism, the focus of his critique is not in the relationship between utopia and ideology per se, but instead in different conceptualizations of architectural utopias. Lefebvre's attempt to come to terms with the problematic coexistence of the utopian and ideological dimension of modernist architecture is hence not based on the rejection of the notion of *utopia* in general, but instead on the attempt to find another conceptualization of utopia moving from reformist conformism to experimentation with different possible modalities of the future.

In the final section, I think through the alternative approach offered by Lefebvre from the point of view of the hypothesis of this book: utopias as methods for social and political change. Rather than residing on the ideologically fixed blueprints, the conceptualization of architectural utopias from this point of view is more interested in experimenting with what is considered possible as a deliberate method for expanding political imagination towards new horizons.

## Modernist Architecture as a Degenerate Utopia?

The interplay between modernist architectural utopianism and its ideological connotations found in the works by Tafuri and Lefebvre deciphers a broader tension related to the role of architecture in the theories of art and politics. Here especially the contested legacy of modernist architectural utopianism serves as an empirical point of reference. Modernism in architecture refers here first and foremost

to architecture's relationship to society rather than to any particular
architectural style. It emerges from the avant-gardist attempt to think
the boundaries between art and life anew. As Hilde Heynen suggests:

> The issues and themes around which the modern movement in
> architecture crystallized are related to the avant-garde logic of
> destruction and construction. Here too what was involved first
> of all was a rejection of the bourgeois culture of philistinism
> that used pretentious ornament and kitsch and which took
> the form of eclecticism. In its stead the desire for purity and
> authenticity was given precedence. All ornamentation was
> regarded as unacceptable; instead, authenticity was required in
> the use of materials, and it was thought that a constructional
> logic should be clearly visible in the formal idiom. In the
> twenties these themes also acquired a distinct political
> dimension: The New Building became associated with the
> desire for a more socially balanced and egalitarian form of
> society in which the ideals of equal rights and emancipation
> would be realized. (Heynen 1999, 28)

How is this then any different from any architectural utopias
given that especially the classical conceptualization of utopias takes
also often the shape of spatial geographical form? Picturing the ideal
city implies always that architecture and city planning are ways to
either produce or maintain ideal social forms and harmony (Pinder
2005). However, while this idea of the ideal city is largely present in
modernist planning and architecture as well, there emerges another
utopian impulse that goes beyond picturing a fixed spatial plan for
an ideal society. Here utopias are considered more as an activity.
By contesting and politicizing the questions associated with the living and spatial structures of everyday life, modernist architecture
directly mounts attacks on previous modes of living and for example classical nineteenth-century bourgeois conceptualizations of
interior and class and the power relations inherent in them. Hence,
this *destructive* aspect of modernism puts forth a different utopian
emphasis towards the classical *construction* of blueprints. Moreover,
the problem with modernist utopianism is no longer the lack of
politics per se (as in classical spatial utopias that imagine worlds
in harmony beyond political conflicts), but the conceptualization of

politics through reified forms of everyday life. This raises a question of whether modern architectural utopias are degenerate, as they do not offer radical negation from society, in the way in which for example utopian painting or literature are believed to offer.

Hence the issue is not only the relationship between architecture and utopia in general, but specifically the particular conceptualization of politics in the utopian aspirations of modernist architecture and the avant-garde. Additionally, this seems to constitute something of a challenge that still defines the future of utopian thought in thinking politically on architectural practices and their utopian aspirations (Henket, Heynen and Allan 2002; Cunningham 2001). But are we talking about ideology or utopia here? Here a good point of reference is the discussion on the convolutions between art and politics within the Frankfurt School and critical theory more broadly. The central common denominator in these discussions is the idea that the utopian dimension of artwork resides in its capability to offer a counter-image, seeing utopia as a negation or counter-image of society rather than deliberately engaging with it. As architecture is necessarily physically rooted in the existing world, however, its utopian dimension appears highly problematic in these readings.

Addressing the relationship between artistic practice and emancipatory political theory and the tendency to consider critical and utopian-dimension artwork to show a different world out-of-reach that is actually only a wish for better being, Gabriel Rockhill, for instance, has shown that the very definition implies the exclusion of architecture. Rockhill maintains that

> regarding design and production, to begin with, architecture
> and public art almost always take place, in our day and age,
> in a constructed milieu, or at the very least within the charted
> territories of traversed landscapes. They cannot, therefore,
> be easily isolated from their immediate inscription in a larger
> sociopolitical space. (Rockhill 2014, 22)

Considered from this point of view, modernist architectural utopias appear as 'degenerate utopias', combining too straightforwardly art and life and being hence examples of how art and politics should not merge (see Coleman 2013).

This pessimistic tone can be traced back especially to Theodor W. Adorno and Ernst Bloch, whose theories on art's utopian capability of showing alternative worlds remain influential. They both suggested, that the core of art's critical and utopian potential is to outline an autonomous expression of a different society, which is in many respects the opposite of, for example, the deliberate attempt to shape society by design having also ideological connotations. These readings have reinforced a certain pessimistic tone towards architecture in proceeding theories of art's role in critical theory defending art's autonomy and considering architecture as unable to propose any real utopian alternative when rather straightforwardly adopting and affirming the capitalist idea of rationalization as the core of modernity.

As already stated, these debates on the relationship between the utopian and the ideological dimension of architecture are primarily about modernist architecture. This is evident especially in Bloch's writings on the utopian function of art. As Nathaniel Coleman argues regarding the possibility to combine Blochian utopian thinking with architectural alternatives, 'Bloch identified the inextricable bond between Utopia and hope in almost everywhere but in architecture' (Coleman 2013, 135). Bloch maintains that modern architecture and its utopian aspiration 'cannot at all flourish in the late capitalist hollow space' given that it is 'far more than the other arts, a social creation' (Bloch 1988, 188). This resonated with Bloch's early work on the utopianism of artwork: For art to be utopian, it needs to have a certain transcendental aura that detaches itself from regular use and everyday objects. Bloch argues that modern architecture with its emphasis on function cannot detach artwork from the current state of things. Instead it is compatible with the modernist ideology of progress creating a 'hollow space of capitalism' (Bloch 1988, 188).

Similarly, with Bloch, Adorno argued that the revolutionary aspect of architecture cannot escape given social relations. Utopian and emancipatory aspirations in modern architecture were according to Adorno 'conditioned by a social antagonism over which the greatest architecture has no power: the same society which developed human productive energies to unimaginable proportions has chained them to conditions of production imposed upon them' (Adorno, cited in Leach 1997, 14). The utopian dimension of art, its ornamental addition that detaches itself from reality, cannot be found in architecture realized through material and social restrictions. These readings seem already to be anticipating that the true nature of the utopia of modernist architecture was somewhat of an ideological compromise which materialized through Fordism and Taylorism (Coleman 2012). This argument is most systematically made by Manfredo Tafuri.

## Utopia Turned Ideology

In 1969, Tafuri published a highly polemic article, 'Toward a Critique of Architectural Ideology', that was later expanded into the book *Architecture and Utopia: Design and Capitalist Development* (1973/1976). Challenging the prevailing art historical explanation of the emergence of modern architecture, Tafuri, informed by Marxist thought, proceeded to see the development of new ideas of modernist architecture and design in a dialectical relationship with capitalist development. The 'utopia' proposed by modernist architects was for him in fact inseparable from the broader capitalist development of the time.

At stake in the criticism of utopianism was the question concerning the nature of the political efficacy of architecture. Rather than revolutionizing the society, the conformist modern architecture, according to Tafuri, offered a reformist way to integrate new social demands into the reorganization of capital. In this way modernist architecture did not present a rupturing counter-image towards the contemporary society or any profound avant-gardist destruction, but instead ended up formulating a 'utopia serving the objectives of the reorganization of production' (Tafuri 1976, 98). This meant constant cross-overs between the Fordist and Taylorist modes of production and an avant-gardist experimental culture in the first part of the twentieth century:

Design, as a method of organizing production more than of configuring objects, did away with the utopian vestiges inherent in the poetics of the avant-gardes. Ideology was no longer superimposed on activity – which was now concrete because it was connected to real cycles of production – but was inherent in the activity itself. (Tafuri 1976, 98)

This can be seen in the context of the broader question concerning reform or revolution in twentieth-century Marxist thought. Here the reformist way implied the realization of utopia in a strange reversed form in which architecture 'becomes a pedagogical act and a means of collective integration' (Tafuri 1976, 132). For Tafuri, the immersion and incorporation of utopian aspects of modernism into broader capitalist development can be understood through the Marxist concept of ideology. In this way Tafuri's reading is undoubtedly informed by (albeit not systematically discussed in relation to) Althusserian structural Marxism on the reproduction of capitalism through various 'ideological state apparatuses' (in this case most importantly, urban planning, housing, architecture, design etc.) that offers a more complex reading of ideologies than just notions of 'false consciousness' (Althusser 2014). In this regard, Tafuri was informed also by Italian autonomist Marxism, including figures such as Antonio Negri, at the same time suggesting that the developments of Keynesian economics, a strong welfare state and Fordist regime of capitalist accumulation are based on somewhat mutual agreement and compromise between capital and (industrial) labor (for Tafuri's political context, see Aureli 2008).

By emphasizing the historical forms of the ideological dimensions of architectural utopias, especially modernism, Tafuri's reading is also to be understood in terms of a 'classical' Marxist critique of utopias, in which the notion of ideology is associated primarily with non-materialist conceptualizations of history such as idealism and its utopian nature. By discussing this failure of non-materialist 'utopian' approaches, especially in relation to architecture and urban planning, Tafuri is hence continuing a critique already familiar from Engels and his fundamental distrust toward other socialist movements and anarchist traditions, but above else the tradition of utopian socialism and its idealistic connotations. In the famous *Zur Wohnungsfrage*, Engels directly confronted this type of utopianism and argued that it provides little more than a bourgeois solution to the housing crisis and that housing is a derivate of broader problems of capitalism that need to be overcome:

> The housing shortage from which the workers and part of the
> petty bourgeoisie suffer in our modern big cities is one of
> the innumerable smaller, secondary evils which result from

the present-day capitalist mode of production. It is not at all a direct result of the exploitation of the worker as worker by the capitalist. This exploitation is the basic evil which the social revolution wants to abolish by abolishing the capitalist mode of production.[1]

This type of Marxist tradition builds itself on the critique of utopianism and its idealistic connotations, which is considered as an opponent to Marxism and historical materialism. The common theme from Engels to Tafuri among others, is a certain detachment between utopian social critique and the analysis of capital and a materialist conceptualization of history. As stated by Tafuri: 'It is useless to propose purely architectural alternatives. The search for alternatives within the structures that condition the very character of architectural design is indeed an obvious contradiction of terms' (Tafuri 1976, 181). Seen this way, utopian traditions not only miss the point, but also to some extent reinforce the ideological dimension of capitalism through its reformist tendencies. In what follows, this contests the whole core of modernist architectural utopianism and the avant-garde: it is literally impossible to change society through design.

However, as I will argue in the next section, this type of conclusion covers primarily utopias as understood from the point of view of their outcomes and different pathways remain if one considers utopias primarily as action and process, immanently revolutionizing what is considered possible (see also Lakkala in this volume). Even though Tafuri's criticism offers a way to highlight the problems related to reformist tendencies of architectural utopias and their lack of emerging critical analysis, especially from the point of view of the critique of ideologies and insights from historical materialism, I maintain that the utopian aim of changing society through design is not necessarily to be interpreted as an attempt to make blueprints, but could also be characterized as a mode of activity: an experimental site for politics. This is based on another conceptualization found in the writings of Henri Lefebvre that is discussed in the following section.

### From Abstract Utopianism to Experimental Utopias

The work by Henri Lefebvre offers another conceptualization of the relationship between modernist architecture, utopias and the coexisting ideological dimension.[2] Additionally Lefebvre, like Tafuri, emphasizes especially the ideological dimension of modernist utopianism. According to Coleman, for example, 'it is precisely the empty promises, false hopes and extravagant failures of modernist architecture and urbanism that preoccupied Lefebvre in much of his writings' (Coleman 2015, 19). However, at the same time Lefebvre never neglects the utopian aspiration of an especially modernist conceptualization of utopias as an activity altogether. Instead, two opposite conceptualizations of a utopia and proponents of utopias exist simultaneously:

> An opposition is continuously at work between abstract and concrete utopias. This enables us to distinguish utopists from utopians ... Abstract utopia relies on technocrats; they are the ones who want to build the perfect city. They concern themselves with the 'real': needs, services, transport, the various subsystems of urban reality, and the urban itself as a system. They want to arrange the pieces of a puzzle to create an ideal. Contrast this with concrete utopia, which is negative. It takes as a strategic hypothesis the negation of the everyday, of work, of the exchange economy. It also denies the State and the primacy of the political. It begins with enjoyment and seeks to conceive of a new space, which can only be based on an architectural project. (Lefebvre 2014, 148)

Here Lefebvre is largely following the distinction between an abstract and a concrete utopia, as outlined by Bloch (see Introduction), but strikingly in the context of urban planning and architecture, areas that Bloch considered generally anti-utopian (see also Coleman 2013; 2015). First, there are negative interpretations associated with (especially modernist) utopianism. Abstract utopianism is frequently referred to as a *scientific and positivistic utopia* by Lefebvre, which can be understood as a conceptualization of utopia that resides only in abstract representations of spaces which dismiss the experimental and transgressive aspects of utopian thinking (Lefebvre 1996, 151). For Lefebvre, this is the most obvious way to think of the spatial dimension of utopias (Lefebvre 2014, 141). Traditional or classical utopias (such as More, Campanella, Bacon) are in Lefebvre's theory 'characterized by an emphasis on architectural form, geometric

ARCHITECTURAL UTOPIAS AND THE (IM)POSSIBLE | **143**

designs and rigid spatial order'. From a more modern perspective, they also 'have a related concern with control, regulation and modes of surveillance' (Pinder 2005, 21).

In this sense, there are important parallels between Lefebvre and Tafuri regarding thinking architectural utopianism and ideology as intertwined. What Tafuri calls architecture as ideology, appears to Lefebvre as a somewhat abstract utopia understood as a conformist engagement and commitment to building a new society. This calls for an urban planning that would integrate new architectural solutions with societal reforms. However, while remaining critical towards general utopianism, and modernist architectural utopias in particular in a similar way to Tafuri, Lefebvre acknowledged that utopias can have a disintegrating role. This leads to an internal distinction between two types of utopias:

> While abstract utopia is a 'positive' extrapolation of the status quo, concrete utopia is 'negative' that is to say it contradicts the premises of the current social order: the everyday defined by the division of labour, economy of exchange, and the state as the primary agent of economic regulation and political subjectivity. (Lefebvre 2014, 151)

For Tafuri, critical architecture practices as such cannot exist, but thinking of architecture, urban planning and housing differently requires an entirely different society and a different economic system. Lefebvre offers a more complex argument regarding the role of relationships between architectural practices and the production of space under capitalism, and hence also a conceptualization of utopia. Encapsulating the difference, Frank Cunningham has suggested that 'like Tafuri, he sees utopian visions as ideologically infused, but they can also serve in an experimental way to prompt challenges to existing structures, functions and forms' (Cunningham 2010, 270). Thinking of utopian as 'a partisan of possibilities' rather than 'utopist', Lefebvre writes:

> For me this term has no pejorative connotations. Since I do not ratify compulsion, norms, rules and regulations; since I put all the emphasis on adaptation; since I refute 'reality', and since for me what is possible is already partly real, I am indeed a utopian;

you will observe that I do not say utopist; but a utopian, yes,
a partisan of possibilities. But then are we not all utopians?
(Lefebvre 1984, 192)

This leads to another conceptualization of utopias found in
Lefebvre's work – to which he simultaneously refers as both a concrete utopia and an experimental utopia. Lefebvre introduced the
concept of an experimental utopia already in 1961 as a critique of
what he called the post-war functional urbanism as the realized legacy of modernist utopianism, where the efficacy of architecture was
primarily understood in terms of how it serves society and repro-
duces the conditions of production. Against this Lefebvre attempted
to find different utopian connotations embodied in architecture as
a category for mediating what is considered possible. For Lefebvre,
this type of utopia is defined as 'explorations of human possibilities, with the help of the image and imagination, accompanied by an
incessant critique and an incessant reference to the given problematic in the real' (Lefebvre 1961; see also Pinder 2015, 37).

Experimental utopia is thus contrasted with abstract utopianism
and the reformist tendencies of modernist architecture. Whereas
abstract utopianism is aimed at providing a certain blueprint for
an ideal society beyond conflicts; the political logic of experimental
utopias seems to be quite the opposite, creating disharmony and
disintegration in the current state of things by expanding the scope
of what is considered possible. To fully understand utopian practice
as a dialectical movement between possible and impossible, blurring
the boundaries between these two categories, Lefebvre maintains
that it is necessary to understand utopias as a sphere for experimentation and invention of new ways of living, thus also disintegrating
the prevailing society and its space. Here utopia is not any grand
ontological statement, but a tactical intervention in the current state
of things and a mediation between the possible and the impossible.
To grasp this idea, Lefebvre used the concept 'transduction', familiar especially from natural sciences, to describe the conversion and
transfer of different constituents:

Transduction elaborates and constructs a theoretical object,
a possible object from information related to reality and a
problematic posed by this reality. Transduction assumes an

incessant feedback between the conceptual framework used and empirical observations. Its theory (methodology), gives shape to certain spontaneous mental operations of the planner, the architect, the sociologist, the politician and the philosopher. It introduces rigour in invention and knowledge in utopia. (Lefebvre 1996, 151)

This approach comes close to what David Harvey, heavily inspired by Lefebvre's theory, has called 'dialectical utopianism' situated in various architectural and spatial practices. It is based on existing concrete social relations and a material basis, but it simultaneously attempts to surpass them within (Harvey 2000). As such, it is an attempt that defines both the work by Lefebvre and Tafuri to show that Marxist thought and utopian thought can coexist. Here a striking example is precisely both the success and the failure of modernist architecture and the classical avant-garde for realizing that every political rupture requires a utopian envisioning of new space:

Change life! Change Society! These ideas lose completely their meaning without producing an appropriate space. A lesson to be learned from soviet constructivists from the 1920s and 30s, and of their failure, is that new social relations demand a new space, and vice-versa. (Lefebvre 1991, 59)

The simultaneous proclaimed failure stated by Lefebvre here is that the picturing of the impossible did not go far enough and was still articulated within the prevailing framework of the possible rather than beyond it. Regardless, the notion of this failure seems to be more of an affirmative critique. Compared to Tafuri, but also to Adorno and Bloch, Lefebvre's theory not only criticizes the history of architectural utopianism for its ideological function of reinforcing the status quo. It also acknowledges the possibility to function also as a form of social criticism by expanding political imagination through architectural forms, experimentations and projects.

Therefore, while remaining critical to the realization of architectural utopianism in the forms Tafuri would call ideology, Lefebvre never neglected the idea of an alternative space beyond capitalism.

It remained a source for inspiration in later forms of experimental architecture. Even though Lefebvre does not exclude the connection between (modernist) architectural utopianism and its relation to capitalist development (as Tafuri does), he argues at least for methodological isolation as

> the only way forward towards clear thinking, the only way to avoid the incessant repetition of the idea that there is nothing to be done, nothing to be thought, because everything is 'blocked', because capitalism rules and co-opts everything, because the 'mode of production' exists as system and totality, to be rejected or accepted in accordance with the principle all or nothing. (Lefebvre 2014, 4)

This is largely echoed by Jameson, who in a similar manner is concerned that a Tafurian-style critique of certain problematic dimensions of particular architectural utopias nullifies the stimulus to think that the surroundings of our lives can be altered.

Here the experimentation related to utopia is understood as a transgressive activity. Going beyond Tafuri, the notion of utopianism of modernism is not rejected in general but its actual outcome is set against its original utopian aspirations. To conclude: compared to Tafuri's critique of utopianism, Lefebvre's understanding of architecture, understood from the point of view of experimental utopia, is not only a reflection of modes of production, but also a site of political struggle and an active ground for reclaiming different meanings, discourses and interpretations.[3]

## Architectural Utopias as Methods for Experimenting with the (Im)Possible

Challenging Tafuri's reading on the ideological dimension of architectural utopias, Lefebvre considered architecture also as a medium for different practices, where collective subjectivities and their relations to political, social and cultural form are opened up for contestation. Lefebvre's solution is hence not to reject utopianism in general, but instead to change the conceptualization of utopias from integration to experimentation. By rejecting the standard wisdom, that architectural utopia necessarily implies a blueprint,

Lefebvre moves the emphasis towards experimental activity when thinking about utopias. This experimental dimension of utopias is understood as transgressive dynamics between the possible and the impossible, which redefines the notion of utopia.

In this sense, configuring the relationship between architecture and utopianism is not just a discussion on how architecture can contribute to thinking about utopian goals. Instead, it is about thinking political transformation itself: What type of act is architectural design? What does it bring to the world? How does it promote a different conceptualization of the future? How does it challenge the existing spatial orders of society? In this sense, architecture not only reflects utopian goals, but constructs them experimentally, bringing future possibilities to the active deliberation of the now.

What is the actuality of these positions for contemporary utopian studies? Based on this assessment, I suggest a re-introduction of Lefebvre's affirmative critique that considers architectural utopias as a sphere for experimenting on what is considered possible as a pathway towards political change. Here utopia is not a fixed outcome, but more of a tool and a method. The methodological notion of utopias is associated especially with the work of Ruth Levitas, who proceeds to consider utopias not as ultimate goals but instead as processes where possible pathways for the future are examined as a methodological way to understand the possibilities for social transformation (Levitas 2013).[4]

This emphasis offers more politically nuanced ways of considering utopias, not only as vague expressions for better being, but first and foremost as tools for political change in various areas of human life. Understanding utopias only as expressions for better being is problematic, in the sense that this makes utopias take distance from society and politics, leaving little room for the actual potentiality of social and political change (Chrostowska and Ingram 2017; see also Lakkala in this volume). After all, especially in the current political climate, it is not only a vague 'utopian wish' that is required, but more concrete, material and effective tools, utopias as methods for and committed to socio-political change even though there are always ideological connotations present in these types of attempts.

Thinking this in terms of how architecture can facilitate this type of social and political change, I have argued throughout this chapter that Lefebvre's theory offers one way to mediate these types of attempts to materialize utopian aspirations. Here again the notion of method is crucial. As Cathy Turner suggests, Lefebvre's somewhat partial isolation from a capitalist mode of production, or what Jameson calls an attempt to claim a semi-autonomous sphere for architecture at least partially separable from its ideological connotation, is first and foremost 'devised to allow or "activate" an alternative or transgressive space in dialectical relationship to established possibilities' (Turner 2015, 4). Here the expected political outcomes are not in a future ideal society, but in the political negotiation of what can be thought.

The conceptualization of utopia in this way is not limited only to Blochian wish-images or a Tafurian critique of ideologies, but it is also embodied in 'real' material forms where 'utopias embrace this tension between dreams and practice. It is grounded in the belief that what is pragmatically possible is not fixed independently of our imaginations, but is itself shaped by our visions' (Wright 2010, 6). The necessary task for thinking of utopias as transformative political action is also to materialize them. This is a process of which architecture as configuring utopian spaces in actually existing forms is a good example. Utopias understood this way, as experimental mediums, aim not at providing a direct spatial setting for a revolution, but this type of utopianism is an immanent, revolutionary action in itself, constantly revolutionizing what is considered possible and the boundaries of the given. The future is the open possibility of becoming something else, utopias are methods for aesthetico-politically configuring that pathway.

## Notes

1  www.marxists.org/archive/marx/works/1872/housing-question/.

2  Lefebvre also confronted Tafuri in person. They both were invited to a conference held in 1972 by the research group on urban sociology located at the Paris 10 University. At the conference, Lefebvre accused Tafuri of having a tendency to explain everything as ultimately working for capitalism thus leaving no room for alternatives (Stanek 2011, 165–167).

3  As Michael Gardiner suggests: 'Lefebvre does not promote a dualistic transcendentalism in which daily life is denigrated, but rather an "everyday utopianism" in which routine and creativity, the trivial and extraordinary, are

viewed as productively intertwined rather than opposed' (Gardiner 2004, 228).

4 Here I am also informed by Nathaniel Coleman, who argues in a similar way, that the notion of utopia as method comes close to Lefebvre's conceptualization of utopia (Coleman 2014; 2015).

## References

Althusser, L. (2014) *On the Reproduction of Capitalism: Ideology and Ideological State Apparatuses*. London: Verso.

Aureli, P. V. (2008) *The Project of Autonomy: Politics and Architecture within and against Capitalism*. New York: Princeton Architectural Press.

Bloch, E. (1988) *The Utopian Function of Art and Literature: Selected Essays*. Cambridge, MA: MIT Press.

Chrostowska, S. D. and Ingram, J. D. (eds.) (2017) *Political Uses of Utopia*. New York: Columbia University Press.

Coleman, N. (2012) 'Utopia and Modern Architecture?' *Architectural Research Quarterly* 16 (4), 339–348.

Coleman, N. (2013) '"Building in Empty Spaces": Is Architecture a "Degenerate Utopia"?' *The Journal of Architecture* 18 (2), 135–166.

Coleman, N. (2014) 'Architecture and Dissidence: Utopia as Method'. *Architecture and Culture* 2 (1), 44–58.

Coleman, N. (2015) *Lefebvre for Architects*. New York: Routledge.

Cunningham, D. (2001) 'Architecture, Utopia and the Futures of the Avant-garde'. *The Journal of Architecture* 6 (2), 169–182.

Cunningham, F. (2010) 'Triangulating Utopia: Benjamin, Lefebvre, Tafuri'. *City* 14 (3), 268–277.

Gardiner, M. (2004) 'Everyday Utopianism: Lefebvre and His Critics.' *Cultural Studies* 18 (2–3), 228–254.

Harvey, D. (2000) *Spaces of Hope*. Edinburgh: Edinburgh University Press.

Henket, H.-J., Heynen, H. and Allan, J. (eds.) (2002) *Back from Utopia: The Challenge of the Modern Movement*. Rotterdam: 010 Publishers.

Heynen, H. (1999) *Architecture and Modernity*. Cambridge, MA: MIT Press.

Hvattum, M. and Hermansen, C. (eds.) (2004) *Tracing Modernity: Manifestations of the Modern in Architecture and the City*. London: Routledge.

Jameson, F. (2000) 'Architecture and the Critique of Ideology'. In K. M. Hayes (ed.) *Architecture Theory since 1968*. Cambridge, MA: MIT Press, 440–462.

Lahiji, N. (ed.) (2012) *The Political Unconscious of Architecture: Re-opening Jameson's Narrative*. London: Routledge.

Lahiji, N. (2019) *An Architecture Manifesto: Critical Reason and Theories of a Failed Practice*. London: Routledge.

Leach, N. (ed.) (1997) *Rethinking Architecture*. London: Routledge.

Lefebvre, H. (1961) 'Utopie expérimentale: Pour un nouvel urbanisme'. *Revue Française Sociologie* 2 (3), 191–198.

Lefebvre, H. (1984) *Everyday Life in the Modern World*. London: Athlone Press.

Lefebvre, H. (1991) *The Production of Space*. Oxford: Blackwell.

Lefebvre, H. (1996) *Writings on Cities*. Oxford: Blackwell.

Lefebvre, H. (2014) *Toward an Architecture of Enjoyment*.

Minneapolis, MN: University of Minnesota Press.

Levitas, R. (2013) *Utopia as Method: The Imaginary Reconstitution of Society*. Basingstoke: Palgrave Macmillan.

Mannheim, K. (1936) *Ideology and Utopia*. London: Routledge.

Margolin, V. (1997) *The Struggle for Utopia*. Chicago, IL: University of Chicago Press.

Miles, M. (2004) *Urban Avant-gardes: Art, Architecture and Change*. New York: Routledge.

Moylan, T. and Baccolini, R. (eds.) (2007) *Utopia Method Vision: The Use Value of Social Dreaming*. Oxford: Peter Lang.

Pinder, D. (2005) *Visions of the City: Utopianism, Power and Politics in Twentieth-Century Urbanism*. New York: Routledge.

Pinder, D. (2015) 'Reconstituting the Possible: Lefebvre, Utopia and the Urban Question'. *International Journal of Urban and Regional Research* 39 (1), 28–45.

Rockhill, G. (2014) 'The Forgotten Political Art *par excellence*? Architecture, Design and the Social Sculpting of the Body Politic'. In N. Lahiji (ed.) *The Missed Encounter of Radical Philosophy with Architecture*. London: Bloomsbury, 19–33.

Stanek, L. (2011) *Henri Lefebvre on Space: Architecture, Urban Research, and the Production of Theory*. Minneapolis, MN: University of Minnesota Press.

Tafuri, M. (1976) *Architecture and Utopia: Design and Capitalist Development*. Cambridge, MA: MIT Press.

Turner, C. (2015) *Dramaturgy and Architecture: Theatre, Utopia and the Built Environment*. New Dramaturgies. Basingstoke: Palgrave Macmillan.

Wright, E. O. (2010) *Envisioning Real Utopias*. New York: Verso.

# 9 | DEMOCRACY AS UTOPIA: ON LOCATING RADICAL ROOTS

*Teppo Eskelinen*

Utopian thought, by its very nature, implies deviation from what immediately exists. The critical departures from the present taken in utopias can come in the form of alternative futures or in the form of alternative spaces, but they are departures all the same. Yet the relation between utopia and reality, as already noted in this book, is more complex than one of total detachment. Utopias can be more or less 'realistic' (Wright 2010), in the sense that their realization seems plausible without the complete undoing of society as we know it; and they can be more or less 'facilitating' (Levitas 2010), in the sense of being relevant for animating transformative social movements and initiatives here and now.

Yet the relation between utopia and reality can also be approached from a third possible direction: locating the utopian potential in what ostensibly already exists. Indeed, a large number of radical and transformative ideas have been partially incorporated in the existing social order, recuperated so that their potential to facilitate further transformation is rendered invisible, and eventually captured in the narrative of the current social order being the superior one with no reasonable existing alternatives.

This chapter analyzes the matter through the notion of democracy. Arguably no other political concept in contemporary society is, simultaneously, as radical *and* compromised. The task is to locate the utopian potential of the concept and see the current hegemonic interpretation of democracy as a hybrid form consisting of both democratic and antidemocratic elements. This analysis also shows, that the currently hegemonic liberal capitalist conception of democracy is merely one possible form of democracy, and indeed quite arbitrary from a historical point of view.

Below, I will discuss the hegemonic interpretation of democracy, understood as compatible with and limited by capitalism and individualistic liberalism. Subsequently, I will sketch what a utopian reading of democracy could look like. Thereafter I will move to discuss mechanisms, both institutional and cultural, that limit democratic ideas and practices today, along with why democratic discourse needs utopias. This leads to a discussion on saving the utopian potential of radical but neutralized concepts and on ways forward with deepening democratic practice. The ultimate aim is to show how democracy can be significantly improved when informed by utopian thought.

## Limited Utopian Concepts

It is fair to say that there exists a given hegemonic democratic practice called liberal capitalist democracy. As noted in the Introduction to this book, this model of democracy is key to the conception of 'the end of history': victorious to the point of distracting human capacity to conceive of qualitatively better societies. While there is a rich tradition of democratic theories (participatory, communitarian, etc.) and different levels of democratic practice (local, global, etc.), one finds this given uniform model of democracy all over the globe. It is based on nation-states as the realm of politics, representation, the popular vote, majority decision-making, constitutionalism, the idea of democracy as mediation between individual rights and government sovereignty, and most importantly the removal of economic matters from the sphere of democracy. While there naturally are minor variations to this model, and one can point to better and worse functioning versions of it, the core model can be said to hold a hegemonic position.

Furthermore, the narrative supporting this hegemonic model sees this particular model as incorporating the spirit of democracy as it has always existed since ancient times. Democracy is then seen as an integral part of the 'liberal capitalist' package, prevailing over grand utopian visions, bringing piecemeal governance and the individual to the forefront in politics.

As always in critiques of hegemony, it needs to be noted how the hegemonic interpretation presents itself as the only viable or natural form, downplaying the possibility of the very existence of other forms. The tendency to see democracy as comfortably aligned with institutions of capitalism and liberalist individualism leads to marginalizing the ostensibly existing variety of 'models of democracy' (Held 2006), to the extent that it can be difficult to even recognize the existence of other variants of democracy. This essentialization also downplays the possibility of the development of democracy: If the form of democracy is seen to be fixed and democracy is seen to be threatened only by totalitarianisms (often associated with utopian aspirations), it is highly difficult to see the shortcomings of the existing model.

The perhaps most defining feature of the hegemonic conception of democracy is its alignment with capitalism. Liberal capitalist democracy sees democracy and the main capitalist institutions as ontologically attached. Therefore, according to this mode of thought, departures from capitalist institutions (private property, commodified labor, markets, the banking system, universal money) would also practically mean departure from democratic practice.

Another specificity of contemporary capitalism is its attachment to liberal individualism. As noted earlier (see Eskelinen, Lakkala and Pyykkönen in this volume; also Jameson 1991 on postmodernism), the hegemonic mentality and the mode of governance more generally today lean strongly towards the individual as the focal point. The idea of the self-interested individuals with incompatible wants then becomes easily also a tacit justification for conservatism in relation to existing institutions; indeed individuality is part of the system of how people are governed today.

Following these starting points, it is convenient to land on the conclusion that the threats democracy is facing today, stem exclusively from a (re-)emergence of totalitarianism, which can be understood as a continuum of twentieth-century totalitarian politics. Essentially, the interpretation of democracy turns into a justification narrative for a degree of conservatism. Certainly, taking this viewpoint, democracy is nothing transformative: It has been achieved and deviations from the existing situation mean deviations for the worse. All one can do to promote democratic values is to defend existing institutions from external threats.

## Utopian Democracy

The alternative to treating democracy as the existing mode of government is to see it as an ideal. As an ideal, it has not been completely institutionalized, and indeed cannot be. The very point of utopia (at least in the non-absolutist sense) is that it can never be fully attained. If, then, democracy is seen in terms of utopia, it will only remain significant if it contains a vision of change. Democracy then is an idea which 'should be conceived as a good that only exists as good so long as it cannot be reached' (Mouffe 2005, 8). This interpretation then sees the threats to democracy as deriving from the loss of its utopian energy, while the liberal interpretation sees utopianism conversely as a threat. Perhaps indeed the best way to even defend democracy as it exists is to try to depart from it to the direction of an ideal.

How could democracy then be understood as an ideal? A straightforward way is to ground democracy on the principle of equality. Or in a stronger formulation, democracy is the antithesis of any form of power based on the superiority of an elite of any kind (Rancière 2009). This means that within the not strictly private domain, all individuals should have an equal influence on the future form of the practices, institutions and principles that constitute this domain. Practically, given the existing hierarchies, mechanisms of exclusion and inequalities in participation, this ideal is quite far from democracy as we know it.

Democracy is, then, 'anyone's power' or political power disregarding identities or personal characteristics. While typically institutionalized in the exclusive form of the nation-state, democracy in the utopian sense can be seen to be the call to empower the human being as such. Democracy is then based on the virtue of humanity as such: Neither acquired nor inherent qualities justify privileged positions. In addition, no power ascending above the community of equals (divinity, nobility, etc.) should have a position to influence the affairs of such community.

Antihierarchy translates organizationally as self-government. The democratic utopia could then be understood as a political community learning how to govern itself without hierarchies. Sometimes this requires emancipation to achieve political subjectivity, for example in independence movements, when the political subject understands that the external power is neither needed for governing nor legitimate. Typically, in contemporary capitalism, what is required is the equalization of political skills through education. In ideal terms, learning self-governance extends to all material positions and identities and becomes a strong anti-elitist stance.

Thinking of 'democracies' as we know them, this very definition of democracy sounds strikingly radical, utopian and revolutionary. These kinds of ideas for the extension of democratic principles are typically labeled 'radical democracy' (e.g. Little and Lloyd 2009), even though it could also be argued that they are rather only true to the idea of democracy, going to 'the root of democracy' (Laclau and Mouffe 2001).

Yet understanding democracy in terms of utopia says nothing about its feasibility, at least in the short term. It is likely that people who are used to hierarchical organizations could not immediately adapt to a democratically organized society. It might be, that many modern institutions would be painstakingly difficult to organize democratically. The ideal therefore should be understood as, first, articulating a direction; and second, insisting on consistency.

Articulating a direction refers to the functions of utopia as showing a point towards which society should be moving, and the rationale of this movement. This point is very illuminatively made by Erik Olin Wright's metaphor of 'the compass' (Wright 2010, Chapter 5), showing the general direction if not the exact path of where the society should be heading. Piecemeal changes then achieve a meaning when set in the context of this general direction. In the context of democracy in contemporary society, this means asking: How are existing hierarchies upheld? What makes the powers of different people to influence social reality so unequal? And further, what kinds of changes could be made to the existing institutions and practices, here and now, to alter the situation?

Insisting on consistency in the context of democracy means that the need for democratic justification knows no fixed or given limits. So, democracy as utopia can be articulated as an extension of principles on the basis of which some existing practices are organized. It is a call to apply democratic principles universally and consistently, understood as a direction for society to develop. There is no reason why democratic politics should out of hand be confined to any pre-defined 'proper' domain, if other domains bear significance to social life, human well-being and relations between people.

Seen this way, democracy is not a form of government, but a principle which can be applied to assess and develop existing practices and institutions, or to imagine completely new ones. Categories such as the market and the state should not be seen as informing pre-determined institutional arrangements. Rather, such categories and institutions, including the language used to describe them, should be kept open to democratic experimentation. Democratic practice should be based on the capacity to see beyond the existing language with its categorizations (see Hietalahti in this volume). This, in the words of Roberto Unger, ideally leads to 'empowered democracy' (Unger 1998). As an illuminating example, C. Douglas Lummis brought attention to the antidemocratic organization of a factory by calling the assembly line an 'antidemocratic machine' (Lummis 1997, 79–110). While perhaps initially outlandish ('how could a machine be political?'), if the functioning of the assembly line depends on a hierarchical order, what is it if not an antidemocratic machine, and isn't the task then to conceive of democratic machines?

## Hegemonic Democracy as a Hybrid Form

What we today call democracies are representative of only one possible constellation, far from being throughout democratic. Yet this does not mean that the hegemonic model of democracy would be an outright hoax: Certainly the ideal has been partially realized. For example, voting, based on the ideas of one vote each, anonymity and the inalienability of voting rights, is a highly democratic practice. Democratic practices also abound in various forms of contentious politics, experimental spaces and microdemocratic initiatives. Yet in other spheres of social organization, quite other principles are applied: For instance, the organization of a government office, a factory assembly line or a competitive market is without question very far from democratic organization.

The existing conception of democracy is more arbitrary than it seems. For example, classical ideas of democracy depart significantly from this contemporary conception, in their conception of the political body and the mode of participation. Furthermore, historical struggles for democracy have not limited themselves to the liberal capitalist conception, by for example demanding voting rights only. From nineteenth-century socialist movements to the South African anti-apartheid struggle prior to the 1990s, the hope invested in the notion of democracy was clearly based on the idea of economic justice (see for instance Slavin 2007 on Rosa Luxemburg; and Congress of the People 1955).

Indeed, all existing 'democracies' are hybrid forms of government, containing both democratic and antidemocratic elements. As a hybrid model, contemporary capitalist liberal democracy constantly seeks balance between the egalitarian ideals of democracy and existing hierarchies. Yet this balance is not an outcome of arbitrary movement, but the capitalist elements of existing society enforce the antidemocratic elements: Therefore, merely protecting existing democratic spaces requires a push to demand a deeper democracy.

But then, looking at these hybrid forms, how exactly is the existing democratic model antidemocratic? The most visible characteristic is the compartmentalization of practices into distinct spheres of 'the political' and 'the economic'. This functions eventually to limit democratic politics into the realm of morality, identity and very small technical issues. Symptomatically, even democratic politics are often assessed from the basis of how politics abide with what is seen as 'neutral' knowledge on society and proper governance of capitalism. As a general notion, the problem related to this distinction was already captured by the old slogan 'democracy is stopped at the factory gate' (e.g. Heller and Feher 1991, 107–110). But the scope of democratic politics is not only limited but can also further be narrowed down, as ever more matters significant to social life and individual well-being take place outside the realm of democratic decision-making. This highlights the essence of liberal capitalist democracy as a paradoxical form of democracy: It simultaneously celebrates the formal democratic procedure, but limits the impact democratic politics can have on society. Ever more people get to vote, but voting has ever less significance (which is not of course to say that it would have zero significance).

Practically, this can be seen in the superior position given to economic institutions. Part of the hegemonic understanding of democracy is that the scope of democratic politics can be limited by international economic agreements (in trade, investment, etc.), which gain a status comparable to a constitution: They are extremely hard to change by democratic means (Gill 1998; Schneiderman 2000). Further, key institutions such as central banks and ministries of finance are increasingly insulated from democratic politics (Mounk 2018; Berman and McNamara 1999). The list could go on, but the point remains: The scope and influence of democratic politics in the existing constellation is quite limited, and could be limited further. Sometimes democratic politics can be close to completely lost to external powers, most noteworthily in the case of 'debtor states', in which the government has entered a debt trap, meaning that external debts accumulate and are not repayable, and creditors, de facto, dictate the policies of the country (e.g. Toussaint 2019).

## Limitations to Democratic Politics

While democracy is kept alive by its utopian manifestations, the liberal capitalist approach to democracy silently justifies its increasing curtailment. Democratic spaces are not stable, but can contract, and indeed holders of economic powers are typically very keen to see that such contraction takes place. If it is seen as legitimate that the experts, rather than a democratic body, decides on, say, central banking, despite its highly political nature, it is easy to apply this form of decision-making to decisions about budgets too. And so forth. If hierarchies are inevitable in organizing production, isn't it natural to apply them in the society more generally? Crucially, justification based on skill is highly different from justification based on democracy.

Antidemocratic ideas have existed throughout history, and there is no reason to believe that they would have suddenly vanished, even though they less often manifest today in the form of trying to deny the popular vote. These ideas can be roughly divided into elitism and belief in expertise. Elites have always perceived the 'masses' as incapable of governing themselves, and the elites themselves as enlightened leaders equipped with the appropriate character to rule. This reflects itself mostly in societies with steep class divisions, but practically everywhere. While elites might be rhetorically for democracy, there remains a mentality that the correct 'dose' of democracy should be carefully administered, so that democracy does not produce unpredictable outcomes and disturb the existing social relations.

Second, contemporary capitalism is an inherently unstable system and requires (undemocratic) expert knowledge to maintain stability. Investors can become frightened by the slightest unanticipated events; multinational enterprises are wary about investing if they do not get binding protections on their investments, overruling future democratic decisions; trade activity needs to be governed by complex sanctioned agreements; and financial markets are filled with securities, as everyone is seeking to get insurance on economic activities. Eventually political change becomes seen as primarily a 'political risk'. As often noted in contemporary criticism, contemporary society is dominated by the idea of technocratic governance based on expert knowledge on society, a 'promise to take politics out of policy' (e.g. Deleon and Martell 2006). As experts possess the skill to administer a highly complex society, they typically see this skill informing also the claim to govern.

Meritocratic expert governance is considerably more successful than is mere authoritarianism. A good example is China, where claims to power are based on effective planning and market freedoms (more so than on brute force). Political legitimation then leans on the skills of the most capable and educated, instead of equality. From the perspective of such meritocracy, democracy is seen as dangerously unpredictable, casting political power to the unenlightened (Bell 2015). The experienced need to govern capitalism predictably in the West and insulating key institutions from democratic influence is indeed closer to the Chinese mindset than Western democrats are happy to admit.

The hegemonic interpretation of democracy, that sees totalitarianism as the only existing threat to democracy, has difficulties in seeing increasing expert power and the concomitant diminishing of the democratic sphere as a threat to democracy. They are neither totalitarian politics nor questioning the formal democratic process. Yet noting these threats to democracy is particularly important as capitalism is developing into an antidemocratic phase: Perhaps contrary to the 1980s, economic dynamism in the contemporary world is associated rather with the lack of democratic practices than their existence. As the connection between democracy and economic growth gets weaker, the true degree of valuation of democracy is revealed. In such circumstances, it is also necessary to be able to articulate the ideal form of democracy.

## The Implications of Liberalism

As noted, another key element of the hegemonic form of democracy is its alignment with liberal individualism. This is reflected in the underlying understanding of the nature of society and democratic practice. In this understanding, society is understood as essentially a system of mediation of the wants and values of individuals existing independently of society. The individual with pre-existing wants and preferences is then taken as the 'entry point' for theory.[1] Thus, social practices and institutions are explained as an outcome of negotiation between these individuals, as they aim to mediate between their conflicting wants. This is where liberal theories of democracy, liberal theories of justice and neoclassical economics coincide.[2]

The most noteworthy issue about individual-based ontology in this context is the difficulty to theorize social change. Seeing society as essentially an arbitrary group of individuals allows sophisticated theorizing on negotiations of different kinds, but it is difficult to analyze, how individuals come together to bring about institutional change, and how people do conceive of alternative kinds of social orders. In the liberal conception of society, politics is essentially seen as a matter of defining the appropriate space for individual rights and for government sovereignty. The prior sphere is where the individual decides based solely on preferences, wants and taste, and has no obligation to give any reasons for given choices. Government, on the other hand, is the sphere of legitimate coercion of existing individual wants.

Democracy, for its part, comes to be seen as essentially a method of conflict mediation. Democracy is then viewed as merely a mechanism of ordering wants, as they cannot all be met, as legislation is crafted and public funds are allocated to various and conflicting purposes. Indeed, as was discussed earlier in this book, politics today is often understood as an exercise in locating individuals within 'value maps' (see Eskelinen, Lakkala and Pyykkönen in this volume). Possible political positions are then reflected as dots in space organized along the axes of right-left and liberal-conservative. This kind of spatial representation of politics ultimately implies that people are ontologically separate and the axes are given.

While this interpretation of democracy does capture relevant aspects of it, other highly important aspects are bracketed by the underlying ontology. When seen as mediation and determining relations between rights and sovereignty, democracy surpasses matters such as learning and development (people can learn horizontal social relations, rather than just pushing their wants), and collective action for social change (human beings are not just individuals with values, but come together to promote new ideas and experiment on them). In other words, the individualist ontology has difficulty recognizing the existence and relevance of utopian ideas and practices. Or, in other words, it is stuck with mediating between 'private hopes', as a deeply democratic sentiment only gets born out of 'public hope' (Lummis 1997, 154–157).

## Open Future and the Significance of Imagination

The definition of ideal democracy can be complemented with one significant underlying notion: open future. This means that in utopian democracy, only the (egalitarian) political community has the power to decide over its future fate. The practices and ideas described above are all obstacles to the ideal of the egalitarian community having the power and responsibility over its future. As the future has become a terrain of political struggle, antidemocratic forces aim at defining the future to the broadest possible extent (expert power), setting priorities that need to be followed in the future as well (e.g. sanctioned investment treaties limiting democratic decision-making), or defining the imaginative framework for future politics (the 'axes' of values). The counter-power to these forces is the democratic spirit, struggling to keep the future open to democratic will against both conservative institutions and conservative knowledge production.

Democracy, by nature, is unpredictable, and indeed a legitimate approach to current democracy is that it is motivated by the need to 'take risk out of democracy' (Carey 1996). Unpredictability emphasizes the aspect of responsibility inherent in democracy. The political body cannot escape deciding its own fate. If the influence of democratic politics is highly restricted, the result is increasing indifference towards politics in general.

As noted before (see Lakkala in this volume), utopias have a function of 'disrupting' the present, which allows seeing a multiplicity of possible open futures. By further providing 'critical counter-images', utopias assist critical reflection on existing society. This is crucial to democracy; a functioning democracy is not only a system of value mediation, but a system of weighing and constructing future alternatives too. This requires that people have the (psychological and cultural) means to construct, compare and critically reflect ideas of what a society could be.

The mere articulation of 'values' is not sufficient, especially if piecemeal change is seen as the exclusive content of politics. Even if politics often functions through such changes, the changes need to be motivated by and connected to broader ideas of a possible society in order to remain significant. If this is not the case, the fear is that democracy is dwarfed into a technical system organized by technical rationality (e.g. Marcuse 1964), with no deeper idea of sense or direction. The technical and formal mode of thought gains dominance over the critical, holistic and reflexive. Politics collapsed into mere governance produces exactly mere technical governance and optimization without articulated purpose (see also Nussbaum 2010 on technical rationality and education). Democratic practice is lost in conjuncture with a broader sense of purpose in society.

If democracy is seen in a broader sense than mediating between ontologically self-contained individuals, political imagination becomes highly important. As noted several times in this book, the political imaginative capacity has been in retreat. In liberal capitalist democracy, not only the possibilities to enact institutional change are limited, but so too are the possibilities to imagine alternative institutions. While technocratization and the lack of political imagination are distinct phenomena, they should be analyzed as a part of the same phenomenon, as elements feeding into a vicious circle. A society of 'capitalist realism' (Fisher 2009) undermines the capacity to imagine alternatives, and conversely, when the capacity to imagine alternatives gets weaker, governance steadily replaces politics. Seen this way, the precondition of a functioning democracy is that people have the proper imaginative skills to conceive of utopias, and capacities to form and develop these products of imagination collectively.

### Saving the Utopian Potential of Concepts and Ways Forward

The project of saving and rearticulating the utopian potential of democracy is an example of a possible utopian strategy within contemporary capitalism. It means to analyze commonplace ideas and ask: Could they still inform hope of radical transformation, despite being diluted, compromised and constrained in the process of their institutionalization? This comes close to what Ruth Levitas called 'the archeological mode' of utopianism: identifying utopian elements in what is otherwise seen as pragmatic or non-utopian (Levitas 2013, 153). Also, one might hear an echo of Ernst Bloch, who sought to find a utopian spirit and expressions of hope in a myriad of contemporary forms of expression (Bloch 1986). Yet the matter with democracy is somewhat different: It is not a pragmatic idea as such, only it has been neutralized. As Wright (2010) tries to map 'real utopias' in the sense of both utopian and achievable, the task here is to seek the utopian in the ostensibly achieved.

Democracy is a brilliant example of a concept which can benefit from this kind of approach. Indeed, despite the apparent radicalism of interpreting democracy as insisting on equality, it has often been noted that democracy rests on the idea of the fundamental equality between human beings, which is deeply inscribed in the Western tradition of political philosophy (Dahl 1989, 84–88). In the case of democracy, mere consistency can inform a radical program.

A key theoretical input in the attempts to rediscover the transformative potential in the liberal use of concepts is Laclau and Mouffe's work on radical democracy. They set out exactly to broaden the remit of liberal democratic principles such as liberty and equality (Laclau and Mouffe 1998). In their words, the problem with 'actually existing' liberal democracies is their incorporated system of power which redefines and limits the operation of those values (Laclau and Mouffe 2001, xv).[3] The task is to separate the values of democracy from the distorting effects of this system of power, described above as consisting of hierarchies, technocratic rule and governance through detached individuality. A large number of similar concepts have a similar kind of potential to be reclaimed. For example, freedom, equality and rights have all carried out substantially more transformative meaning, but then been incorporated into the existing framework and lost their utopian potential, at least to a large extent (see e.g. McNeilly 2016 on radical interpretations of human rights).

Seeing achieved democratic progress assists in seeing the present as a point in a continuum between absolute hierarchy and ideal democracy. Indeed, when looking at the history of democracy, highly pivotal democratic developments have taken place within democratic revolutions such as the end of the Cold War. This applies also culturally: Still not so long ago, the 'layman' was seen as essentially unfit for democratic decision-making. The universalization of education systems has had a remarkable effect on the equalization of democratic participation. Also, there has been a remarkable change in gender relations, largely undoing the situation in which half of humanity was blocked from political participation. From the perspective of society a hundred years ago, current democracy would have appeared as a risky attempt to give power to the unfit. There is no reason why further steps in the direction of 'anyone's governance' could not be taken.

Promisingly, new initiatives have recently revitalized a discourse on developing democracy within countries that used to see themselves as self-evidently democratic. These include, for example, participatory decision-making mechanisms such as participatory budgeting (Baiocchi 2001) and direct democracy (Kaufmann, Büchi

164 | T. ESKELINEN

and Braun 2010), along with global democracy initiatives (Teivainen and Patomäki 2004) and experiments (in the World Social Forum as an experiment in global and participatory democracy, see Smith et al., 2014). Yet even perhaps more importantly, new movements have highlighted issues related to the common ownership of resources (McLaren and Agyeman 2015). More generally, the distinction between democracy and economy is continuously contested, by for example showing how economic exclusion leads to democratic exclusion (McLaren and Agyeman 2016, 76–77).

Sometimes new democratic initiatives stem from new economic forms such as the 'solidarity economy'. For example, Ethan Miller asks, 'Could a process of horizontal networking, linking diverse democratic alternatives and social change organizations together in webs of mutual recognition and support, generate a social movement and economic vision capable of challenging the global capitalist order?' (Miller 2019). All in all, current (counter-)political discourse is rich with initiatives with significance to democracy. Yet most of such initiatives relate to grassroots organization or contained spaces (for an exception see Albert and Hahnel 1991 on participatory economics). The challenge is then, how to diffuse the new practices, pedagogies and ontologies experimented with in these contained spaces to inform broader institutional developments.

## Conclusions

This book began with a complaint on the existing state of utopian thought. Current liberal democratic capitalism involves a number of institutions, practices and mindsets that, along with the general cultural mood, significantly curtail imagination on societal alternatives. As noted in this chapter, one of the mechanisms of this curtailment is the dilution of radical concepts. While people have long fought for democracy, liberty, freedom of speech and so forth, liberal capitalism materializes some limited aspects of these ideals and presents itself as the ultimate incarnation of these ideals. Therefore, a possible approach for formulating political utopias is to try to 're-radicalize' such concepts which have been subject to recuperation, rather than distancing oneself completely from the vocabulary dominating current politics.

Above, I have discussed a notion of radical (or consistent) democracy, in order to show how widely accepted democratic principles could lead to highly radical conclusions. Except for radical departures from the existing order, it is important to see utopian articulations of democracy necessary for even the continued existence of democracy as we know it, as the democratic space is always under pressure. This pressure derives from the system of power consisting of economic interests and technocratic governance, and the interpretation of democracy as a mediation space between ontologically unattached individuals. Democracy has to be on the move towards deeper democracy, articulated as a motivating but unattainable ideal; or, in other words, 'the horizon for democracy is the impossibility of the full realization of democracy itself' (McNeilly 2016).

The energy for democracy comes from the quest for transitions to equality, and conversely, when democracy declares itself 'achieved', it ceases to be in motion, loses its utopian spirit and goes into retreat. Furthermore, democratic practice is dependent on (collective) learning, communication, deliberation and public criticism. Without a utopian sentiment, democratic deliberation becomes devoid of substantial content. Yet while democracy is far from being realized in the currently existing self-declared democracies, it can be seen as something of a 'limited realized utopia'. While strong societal forces currently curb its further realization, contemporary hybrid democracy clearly has democratic spaces and practices. This leads to a need to see the radical and utopian elements in the self-evident, given and widely accepted. As noted above, this approach can be taken as a method to analyze other concepts as well from the perspective of rearticulating their transformative potential.

## Notes

1  All descriptions of social reality need an entry point, consisting of the basic ontological and methodological approach. Theoretical entry points cannot be assessed, so that ultimately a suitable method could be used to determine the superior entry point; rather they merely can be assumed, and the best one can do is to be aware of the consequences of this choice.

2  Illuminatingly, for instance polling techniques used to predict electoral outcomes were invented for the purposes of justifying marketing, promoting the idea that each individual has hidden wants that need to be brought to knowledge (Lears 1995).

3  Not everyone shares this optimism towards 'rescuing concepts'. Rancière, for example, has been highly skeptical of attempts to maintain liberalism or its resources in any form (Chambers 2013, 10–14).

# References

Albert, M. and Hahnel, R. (1991) *The Political Economy of Participatory Economics*. Princeton, NJ: Princeton University Press.

Baiocchi, G. (2001) 'The Porto Alegre Experiment and Deliberative Democratic Theory'. *Politics & Society* 29, 43–72.

Bell, D. (2015) *The China Model: Political Meritocracy and the Limits of Democracy*. Princeton, NJ: Princeton University Press.

Berman, S. and McNamara, K. (1999) 'Bank on Democracy: Why Central Banks Need Public Oversight'. *Foreign Affairs* 78 (1), 2–8.

Bloch, E. (1986) *The Principle of Hope*, Volume 1. London: Blackwell.

Carey, A. (1996) *Taking the Risk out of Democracy*. Chicago, IL: University of Illinois Press.

Chambers, S. (2013) *The Lessons of Rancière*. Oxford: Oxford University Press.

Congress of the People (1955) *The Freedom Charter*. Retrieved from www.historicalpapers.wits.ac.za/inventories/inv_pdfo/AD1137/AD1137-Ea6-1-001-jpeg.pdf.

Dahl, R. (1989) *Democracy and Its Critics*. New Haven, CT: Yale University Press.

Deleon, P. and Martell, C. (2006) 'The Policy Sciences: Past, Present and Future'. In B. Peters and J. Pierre (eds.) *Handbook of Public Policy*. Thousand Oaks, CA: Sage, 31–48.

Fisher, M. (2009) *Capitalist Realism: Is There No Alternative?* Winchester and Washington, DC: O Books.

Gill, S. (1998) 'New Constitutionalism, Democratisation and Global Political Economy'. *Pacifica Review: Peace, Security and Global Change* 10 (1), 23–38.

Held, D. (2006) *Models of Democracy*. Cambridge: Polity Press.

Heller, A. and Feher, F. (1991) *The Grandeur and Twilight of Radical Universalism*. New Brunswick, NJ: Transaction Publishers.

Jameson, F. (1991) *Postmodernism, or, the Cultural Logic of Late Capitalism*. Durham, NC: Duke University Press.

Kaufmann, B., Büchi, R. and Braun, N. (2010) *Guidebook to Direct Democracy: In Switzerland and Beyond*. Marburg: Initiative and Referendum Institute Europe.

Laclau, E. and Mouffe, C. (1998) 'Hegemony and Socialism: An Interview with Chantal Mouffe and Ernesto Laclau'. *Palinurus*. Retrieved from http://anselmocarranco.tripod.com/id68.html.

Laclau, E. and Mouffe, C. (2001) *Hegemony and Socialist Strategy: Towards a Radical Democratic Politics*. 2nd Edition. London: Verso.

Lears, J. (1995) *Fables of Abundance: A Cultural History of Advertising in America*. New York: Basic Books.

Levitas, R. (2010) *The Concept of Utopia*. Bern: Peter Lang.

Little, A. and Lloyd, M. (2009) *The Politics of Radical Democracy*. Edinburgh: Edinburgh University Press.

Lummis, C. D. (1997) *Radical Democracy*. Ithaca, NY: Cornell University Press.

Marcuse, H. (1964) *One-dimensional Man: Studies in the Ideology of Advanced Industrial Society*. London: Routledge & Kegan Paul.

McLaren, D. and Agyeman, J. (2015) *Sharing Cities*. Cambridge MA: MIT Press.

McLaren, D. and Agyeman, J. (2016) 'Sharing Cities: Governing the City as Commons'. In J. Ramos (ed.) *City as Commons: A Policy Reader*. Melbourne: Commons Transition Coalition, 77–79.

McNeilly, K. (2016) 'After the Critique of Rights: For a Radical Democratic Theory and Practice of Human Rights'. *Law and Critique* 27 (3), 269–288.

Miller, E. (2019) *'Other Economies Are Possible!': Building a Solidarity Economy*. Retrieved from www.geo.coop/node/35.

Mouffe, C. (2005) *The Return of the Political*. London: Verso.

Mounk, Y. (2018) 'The Undemocratic Dilemma'. *Journal of Democracy* 29 (2), 98–112.

Nussbaum, M. (2010) *Not for Profit: Why Democracy Needs the Humanities*. Princeton, NJ: Princeton University Press.

Rancière, J. (2009) *Hatred of Democracy*. London: Verso.

Schneiderman, D. (2000) 'Investment Rules and the New Constitutionalism'. *Law & Social Inquiry* 25 (3), 757–787.

Slavin, P. (2007) 'Rosa Luxemburg's Concept of Democracy'. *Proceedings of the 15th International Rosa Luxemburg Conference*. Tokyo: Chuo University. Retrieved from http://www2.chuo-u.ac.jp/houbun/sympo/rosa_confe2007/pdf/papers/Slavin.pdf.

Smith, J. et al. (2014) *Global Democracy and the World Social Forums*. Boulder, CO and London: Paradigm Publishers.

Teivainen, T. and Patomäki, H. (2004) *A Possible World: Democratic Transformation of Global Institutions*. London: Zed Books.

Toussaint, E. (2019) *The Debt System: A History of Sovereign Debts and Their Repudiation*. Chicago, IL: Haymarket.

Unger, R. M. (1998) *Democracy Realized: The Progressive Alternative*. London: Verso.

Wright, E. O. (2010) *Envisioning Real Utopias*. London and New York: Verso.

# INDEX